Adoptive Management Innovation

Haifen Lin

Adoptive Management Innovation

Springer

Haifen Lin
Faculty of Management and Economics
Dalian University of Technology
Dalian
China

ISBN 978-981-10-7711-1 ISBN 978-981-10-7712-8 (eBook)
https://doi.org/10.1007/978-981-10-7712-8

Library of Congress Control Number: 2018933501

Printed on acid-free paper

This Springer imprint is published by the registered company Springer Nature Singapore Pte Ltd.
part of Springer Nature
The registered company address is: 152 Beach Road, #21-01/04 Gateway East, Singapore 189721, Singapore

Preface

As competition intensifies and globalization accelerates, innovation has been considered as a main source of competitive advantage and economic growth. Although there are abundant academic research and writings on innovation, most focus on technological innovation and its potential for producing explicit and significant benefits. This was until Stata argued that the bottleneck of many U.S. companies was management innovation, rather than technological innovation. Management innovation is particularly prerequisites and facilitates efficient use of technical products and process innovations, thus improving organizational performance through productivity, lead times, quality, and flexibility. Hence, it has become one of the most important and sustainable sources of competitive advantage. Henceforward, management innovation has attracted extensive attention to realize its critical contributions to long-term success of firms.

Literatures present that there are two types of management innovation: one refers to a practice or a structure, which is new to the state of the art without known precedent and named as generating innovation; and the other one refers to something that is novel to the firm and is adopted from another context, which is named as adoptive management innovation. Though the former enables production of more new knowledge, the latter has been more popular among firms in China and perhaps around the whole world. It is because the basic purpose of management innovation is to utilize organizational resources more efficiently and further their goals with little intention to pursue differentiation and without any protection from the patent. In addition, the introduction and implementation of management innovation that is novel to the organization is also of high uncertainty. The success of new practices may rely on their adaptation to their idiosyncratic context within the organization where they are adopted, which indicates a high value in focusing on this body of innovations.

Though researchers have paid much attention to why management innovation matters, what antecedents may affect innovations, how new management ideas or methods are delivered, and what conditions give rise to the emergence and diffusion of management innovation. Surprisingly, seldom research has gone into exploring how more popular adoptive management innovation occurs, how managers in a firm make the decision of adopting new practices, why management innovation so often fails to yield intended results, or how management innovation affects organizations. Considering the importance and popularity of adoptive management innovation in China and perhaps around the world, this research attempts to focus on the process, adoption decision, driver, and effects of adoptive management innovation, aiming to address literature gaps and offer implications for managerial practice, by focusing on five issues about adoptive management innovation.

First, what is the process through which adoptive management innovation occurs? Based on the conception of what makes adoptive management innovation unique, this book attempts to investigate the case of the Organizational Efficiency Management of Jiangxi Mobile in China to identify key activities of adoptive management innovation and develop a two-interlinked-subprocess framework of adoption decision and implementation, aiming to offer suggestions for firms in adopting new management practices. The results indicate that adoption of existing management practices or methods from somewhere else is a complex and logical process rather than a simple one of knowledge transferring. It needs to integrate existing practices into new organizational context and establish their innovative value during implementation. One core element of this process framework is the emphasis on activities of problem diagnoses and realization of the fitness between management practices adopted and the new organizational context, and another one is the sequence of activities in the whole process.

Second, how do core managers of a firm make the decision of adopting new management practices to improving its efficiency? Based on exploration on the decision process of adoptive management innovation and responsibility of managers in each stage, this research further extracts three major activities of managers in the process of adoption decision, referring to innovation intention, knowledge acquirement, and risk perception, and then establishes a three-dimensional decision framework made up by paths from entrepreneurial orientation to innovation intention, social network to knowledge acquirement, and cognitive biases to risk perception by further combining with relevant literatures and decision practices. Data on 237 managers from different firms were collected to examine the model by adopting the method of structural equation modeling, with the purpose of showing affecting paths and internal mechanism of complex decision-making. The results indicate that innovation intention, knowledge acquirement, and risk perception directly affect management innovation decision level; innovation intention, knowledge acquirement, and risk perception are influenced by entrepreneurial orientation, social network, and cognitive biases of managers, respectively; besides the three relevantly independent affecting paths, interrelations also exist across paths; the three main factors may produce both direct and indirect effects.

Third, how could management innovation be implemented deeply at the most microlevel of organizations, namely organizational routines? By adopting a longitudinal interpretive and exploratory case study on the case of Day-Definite (DD) innovation which successfully brought Arima World Group Company Limited (HOAU) into a new value-added arena, in terms of timing, security, and high service quality, this research attempts to explore the complex implementation mechanism of management innovation. Multiple approaches were utilized during data collection to meet criteria for trustworthiness, including semi-structured interviews, archival data, and observation; and the data analysis went through a five-step process. The results confirm management innovation a complex project concerning organizational routines which represent a central and fundamental element of organizations. Also, it finds that organizational routines evolve in innovation implementation through a three-phase process consisting of the existing-routine-domination phase, the new-routine-creation phase, and the new-routine-solidification phase, each exhibiting different innovation activities and characteristics of participants' cognition and behaviors; recreation of new routines is the key for routine evolution, thus for success of management innovation.

Fourth, how could adoptive management innovation be internally driven by dynamic capabilities? Besides external pressure, leadership behaviors, and interaction between context and search, management innovation relies on the whole organizational system with valuable, rare, inimitable, and non-substitutable resources and especially capabilities of integrating, reconfiguring, gaining, and releasing resources. Before examining the effects of dynamic capability, a measuring scale for adoptive management innovation was set up by adopting a process-oriented method rather than a result-oriented one, which may further research on adoptive management innovation and offer valuable insights for measuring generating innovation and other complicated organization behaviors. PLS-SEM was adopted to examine both the structural model reflecting the relationships between four components of dynamic capabilities and four phases of adoptive management innovation (which was reset up by holding an organizational perspective in Chap. 5) and the measurement model reflecting indicators of each construct. The result reveals that each component affects each phase with a unique path, and relational capabilities based on relationships with firms and individuals serves as a fundamental role, which reflects the culture in China where affection rooting in relationships determines acquisition of resources and even survival and development of organizations.

Fifth, how could intangible (adoptive) management innovation with ambiguous outcomes affect organizations? Considering that management innovation is typically tacit in nature and full of uncertainty and ambiguity, this research argues that it may contribute to organizations by directly serving tangible product innovation, especially complex products with a separate brand. A general framework that explicated support and promotion mechanism of management innovation to the process of complex product innovation was set up through a longitudinal interpretive and exploratory case study on Lexus of Toyota Motor. The results indicate that both explicit management innovations (i.e., observable management practices) and implicit management innovations (i.e., unobservable management concepts) may interdependently coexist in the firm when creating a complex product with a separate brand, and the former may support the process of product innovation through the latter.

Though this research may offer various implications to both theory and managerial practice, it is constrained by four *limitations at least*, which also represent fertile ground for future research in this area. First, the findings of the two-subprocess framework of adoptive management innovation, the decision-making process, the evolution of organizational routines in innovation implementation, and the effects of management innovation on performance of an organization need to be further confirmed for more organizations. That is, more cases of adoptive management innovation should be adopted and compared to improve these frameworks and further explore into details. Second, the research does not investigate other variables that may affect innovation process, innovation decision, and innovation implementation, such as contextual elements like competitive intensity, internal elements like structure and culture, as well as top management teams. Third, though this research has focused on four components of dynamic capability, there may be some others not mentioned, e.g., seizing capability, adaptive capability, or learning capability. Additionally, the data were collected by questionnaires, so further longitudinal research using observation and interviews could contribute to this area by providing a richer understanding.

In addition, this research does not address other relevant problems, e.g., why some managers pursue innovative opportunities by introducing new management practices while others do not; how dynamic capability affects the process of generating management innovation; how resources of firms which service as the base for dynamic capability affect the formation of dynamic capability and even management innovation; how product innovation affects management innovation in a firm; or how other elements of firms (e.g., competitive intensity, structure, culture, or top management teams) affect the process of management innovation, especially in the context of China. Therefore, future research may be concerned with these issues.

Keywords: Adoptive management innovation • Innovation process
Adoption decision • Innovation implementation • Innovation driver
Innovation effects

Dalian, China Haifen Lin

Acknowledgements

The author is supported by the National Natural Science Foundation of China Project (71572025, 71372082, 71632004), Project of key research base for Humanities and social sciences of Ministry of Education (15JJD630004). The author is grateful to Angela Higgins, Jingqin Su, Mengya Chen, and Ren Shang for helping improve this research. The author also appreciates the guidance and support of editors from Springer.

Contents

About the Author

Haifcn Lin is an Associate Professor of Strategic Management at the Faculty of Management and Economics, Dalian University of Technology. She received her Ph.D. from Dalian University of Technology. Her research focuses on management innovation and organizational routines.

Chapter 1
Introduction

1.1 Research Background

Under the conditions of increased global competition, technological changes, fast-changing market situations and continuous customer demand for quality services, innovation has been regarded as a main source of competitive advantage and economic growth (Ekvall and Arvonen 1994; Howell and Higgins 1990; Tushman and O'Reilly 2002; Damanpour and Schneider 2006). As a stream of action involving a series of activities that create and diffuse new useful products, innovation is motivated by opportunities for value creation resulting from novel technologies, new needs, fresh links between existing technologies and needs, or tensions and gaps between various components of current systems (Dahmén 1970; Hughes 1983; Schumpeter 1950). Schumpeter first emphasized the positive impact of innovation on economic development and defined innovation in his famous work The Theory of Economic Development in 1912 as new combinations of new products, new production methods, new forms of organization, new markets, and new sources of supply. Henceforward, innovation research has attracted extensive attention from both scholars and practitioners and become a hot research theme over almost a hundred years.

Schumpeter's definition of innovation lays particular stress on technological innovation. Accordingly, driven by the innovation theory from Schumpeter, most of the existing vast bodies of academic researches and writings on innovation focus on technological innovation (e.g., Henderson and Clark 1990; Utterback 1994) and ignore more implicit innovation approaches within organizations, such as management innovation, service innovation and strategic innovation, etc. Until the late 1980s, Ray Stata, the former CEO of Analog Devices Inc., a semiconductor company located in Norwood, Massachusetts, argued that the bottleneck to progress of many U.S. companies was not product or process innovation but management innovation, and considered management innovation as a necessary condition of fully using the advantage of technology, which broke the log jam of the dominance of technological

H. Lin, *Adoptive Management Innovation*,
https://doi.org/10.1007/978-981-10-7712-8_1

innovation (Birkinshaw et al. 2008). Since then, more and more researchers have directed their attention towards management innovation to realize its critical contributions to long-term firm success (e.g. Birkinshaw and Mol 2006; Mol and Birkinshaw 2009; Birkinshaw et al. 2007, 2008; Hamel 2006, 2007; Vaccaro et al. 2012; Damanpour et al. 2009; Lazonick and Teece 2010). The reason why the theoretical circle focuses on management innovation is the rise and prevalence of management innovation practices (e.g., the Japanese enterprises have been able to rise rapidly and catch up with the strong competitors in Europe and the United States after the second world war resulting from the new management practices of Total Quality Management and Lean Production; Haier has become the head of the domestic industry through management innovation such as "market chain" business process reengineering, and successfully went to global market, which make great encouragement to a large number of enterprises to take the path of management innovation). In fact, over the past century, global major events relevant to management innovation occurring in different periods have caused a stir among managers world widely and waves of management changes, and made great contribution to the development of the whole management discipline and the promotion of management efficiency.

In recent years, as the market increasingly homogenized and product life cycle shortened, differentiated competitive advantages resulting from technology, business and service gradually became weakening, thus many companies with weaker research strengths can't get significant advantages through technological innovation, while management innovation has become new competitive advantages of organizations by fundamentally addressing deep-seated problems of low organizational efficiency, backward productivity and organizational inflexibility so as to improve organizational efficiency, performance and efficient use of resources (Hamel 2006), through which forms an important support to technology or product innovation from organization changes (Stata 1989). A management innovation creates long-lasting advantages when it meets one or more of three conditions: the innovation is based on a novel principle that challenges management orthodoxy; it is systemic, encompassing a range of processes and methods; and it is part of an ongoing program of invention, where progress compounds over time. Hamel (2006, 2007) in particular argues forcefully that in today's age management innovation may represent one of the most important and sustainable sources of competitive advantages for firms because of its context specific nature among others, or "Over the past 100 years, management innovation, more than any other kind of innovation, has allowed companies to cross new performance thresholds" (Hamel 2006, p. 73). This makes any study into this topic particularly relevant for practice but also important from the perspective of the study of sustainable competitive advantage, a key domain of strategic management and other academic areas (Mol and Birkinshow 2009). However, among various management innovation practices, except for a few successful cases, most enterprises not only failed to achieve the desired effect but fell into a dilemma after the introduction and implementation of new practices. Obviously, the systematizing, high risk and complexity of management innovation result in the high difficulty of innovation practices, and the particularity and difference of the internal and external context of the organization can easily lead to the failure of new management practices. Therefore, it is urgent to improve the effects of introduction and implementation of the management innovation.

Based on relevant literature, Mol and Birkinshow (2009) present that there are two types of management innovation: one refers to a management practice or a particular structure, which is all new to the state of the art without known precedent and named as generating innovation (e.g., Toyota Motor Corp.'s Lean Production System and Procter and Gamble Corp.'s Brand Management Model) (Birkinshaw et al. 2008); and the other one refers to something existing already that is novel to the firm and is integrated and adopted from another context, which is named as adoptive management innovation (e.g., Vaccaro et al. 2012; Su and Lin 2010; Lin and Su 2010). For example, McDonald's has been imitated widely by many restaurants and chain management enterprises around the world after it started a standardized business model; similarly, General Motors completed the transformation of the original organization to the division structure in the early 1920s which was called the paragon of division structure. Though the former enables production of more new knowledge, the latter has been adopted by more firms in China and even the whole world. It is because the basic purpose of management innovation is to utilize organizational resources more efficiently and further their goals with little intention to pursue differentiation and without any protection from the patent (Teece 1980). In addition, the introduction and implementation of management innovation that is novel to the organization is also of high uncertainty. The success of new practices may rely on their adaptation to their idiosyncratic context within the organization where they are adopted (Ansari et al. 2010; Vaccaro et al. 2012), which indicates a high value in focusing on this body of innovations. That is to say, from the nature of management innovation, even the adoptive management innovation requires the organization to make radical changes to the routines or genes (Birkinshaw et al. 2008) and achieve the adaptation between the new practices and the internal and external context of the organization.

However, the literature discussing adoptive management innovation contains gaps. Among the existing literatures, studies on management innovation process lay particular stress on the generating innovation, while research on innovation spread usually focuses on adoptive management innovation, introducing the existing management practices into a new organization or domain. Despite the fact that there is almost no distinction between both two types, both are listed as research objects in most cases by other researchers. Though many studies on management innovation focus on generating type (e.g. Kimberly and Evanisko 1981; Birkinshaw and Mol 2006; Birkinshaw et al. 2007, 2008) or blend the two types together (e.g. McCabe 2002; Howell and Higgins 1990), some others have realized the importance of exploring adoptive management innovation. For example, proponents of innovation diffusion perspective pay attention to how existing practices are delivered (e.g. Abrahamson 1996; Burns and Wholey 1993; Teece 1980). Vaccaro et al. (2012) focus on management innovation at the organization level and investigate the role of leadership behavior as a key antecedent; Mol and Birkinshaw (2009) demonstrate a trade-off between context and search, where there is a negative effect on adoptive management innovation associated with their joint occurrence. In a word, researchers have paid much attention to why management innovation matters (e.g. Hamel 2006), what antecedents may affect innovations (e.g. Mol and Birkinshaw 2009; Vaccaro et al. 2012), how new management ideas or methods are delivered (e.g. Abrahamson 1996; Teece 1980) and what conditions give rise to the emergence and diffusion of management innovation (e.g. Guillén 1994; Kossek 1987). Surprisingly, seldom research has gone into exploring how more popular adoptive management innovation occurs, while that of the generating type of innovation has been clearly illustrated by Birkinshaw and Mol 2006, Birkinshaw et al. (2007, 2008), why management innovation so often fails to yield intended results or how management innovation affect organizations. From the perspective of the macro realization mechanism and the micro-process mechanism of the adoptive management innovation, this book attempts to address these literature gaps and offer implications for management practices, expecting to provide guidance for various firms to introduce and implement management innovation practices.

1.2 Research Contributions

This book focuses on adoptive management innovation and explores different phases through which adoptive management innovation occurs, how managers of a firm make the decision of adopting new management practices, how firms implement innovation deeply into the most micro level of organizational routines, how dynamic capability of an organization drives the whole process of adoptive management innovation and how management innovation affects organizations through tangible product innovation, so as to address both theoretical and practical gaps by considering five specific issues.

First, as for adoptive management innovation and generating innovation, which does need more attention, especially in China? An investigation into the practice of Chinese firms indicates that a large number of firms have been rushing into the wave of management innovation by adopting advanced management practices that have already been successfully implemented elsewhere especially western developed companies, with few new-to-the-state-of-the-art management practices created. Similarly, firms around the world prefer to adopt existing management practices instead of creating a new method, due to its freedom of delivery and low cost of adoption. Therefore, in line with that of Vaccaro et al. (2012), Mol and Birkinshaw (2009) and Lin and Su (2014), this research investigates management innovation at the organizational level of analysis by focusing on management innovation that is new to the firm.

Second, how does adoptive management innovation occur? Since the nature of adoptive management innovation is different from that of generative innovation, it may go through a different process from that of new-to-the-state-of-the-art management innovation proposed by Birkinshaw and Mol (2006), Birkinshaw et al. (2008). Furthermore, adoption means advancing a new scheme by integrating existing practices into problems and resources of the new organization (Mol and Birkinshaw 2009) rather than simply imitating or repeating practices of other organizations. That is, it is a complicate process needing to be further explored, especially in the special context of China where firms are not developed enough to satisfy the demand of existing advanced management practices. In a word, this research aims to address how management practices implemented successfully elsewhere have been introduced into less-developed Chinese firms and implemented effectively in Chap. 3, or to investigate key activities of the generative mechanism through which an adoptive management innovation occurs.

Third, how core managers of a firm make the decision of adopting new management practices to improving its efficiency? As a big issue concerning the overall benefit of organizations as well as their future development, whether to introduce a new management practice or method depends on core managers (Elenkov et al. 2005; Kimberly and Evanisko 1981) who take the responsibility of directing future development of organizations by making essential decisions. Why organizations in the same industry or surrounded by similar environment seldom take same actions of management innovation? The difference of managers in decision making concerns a lot. However, what main actions do managers take during the process of decision making? Or what are the causes for the difference? In order to address these questions, this research attempts to establish a three-dimension decision model based on an exploration on decision process of adoptive management innovation and responsibilities of managers in each stage, and further examines the model through Structural Equation Modeling (SEM) on the data from 237 managers in Chap. 4, with the purpose of uncovering affecting paths and internal mechanism of complex decision making. The results indicate that innovation intention, knowledge acquirement and risk perception directly affect management innovation decision level; innovation intention, knowledge acquirement and risk perception are influenced by entrepreneurial orientation, social network and cognitive biases of managers respectively; besides the three relevantly independent affecting paths, interrelations also exist cross paths; the three main factors produce both direct and indirect effects.

Fourth, how could management innovation be implemented deeply at the most micro level of organizations, namely organizational routines? Management innovation potentially requires fundamental changes in the routines of an organization, which reflects the difficulty and complication of innovation implementation (Argyris and Schän 1978; Birkinshaw et al. 2008). Organizational routines, defined as repetitive, recognizable patterns of interdependent actions involving multiple actors (Feldman and Pentland 2003, p. 96; Howard-Grenville 2005, p. 618; Parmigiani and Howard-Grenville 2011, p. 417), have been regarded as a basic unit of analysis for organizational innovation and even a source of innovation (Feldman and Pentland 2003). By adopting a longitudinal interpretive and exploratory case study on the case of Day-Definite (DD) innovation which successfully brought Arima World Group Company Limited (HOAU) into a new value-added arena, in terms of timing, security, and high service quality, this research, this research attempts to explore the complex implementation mechanism of management innovation. Multiple approaches were utilized during data collection to meet criteria for trustworthiness, including semi-structured interviews, archival data, and observation; and the data analysis went through a five-step process. The results confirm that management innovation is a complex project concerning organizational routines which represent a central and fundamental element of organizations. Also, it finds that organizational routines evolve in innovation implementation through a three-phase process consisting of the existing-routine-domination phase, the new-routine-creation phase and the new-routine-solidification phase, each exhibiting different innovation activities and characteristics of participants' cognition and behaviors; recreation of new routines is the key for routine evolution, thus for success of management innovation.

Fifth, how adoptive management innovation could be internally driven by dynamic capability? Or how dynamic capability could efficiently enhance the process performance of adoptive management innovation? Besides external pressure (DiMaggio and Powell 1983), leadership behaviors (Vaccaro et al. 2012) and interaction between context and search (Mol and Birkinshaw 2009), the whole organizational system with valuable, rare, inimitable, and non-substitutable resources (Barney 1991; Teece et al. 1997; Wernerfelt 1984) and especially capabilities of integrating, reconfiguring, gaining and releasing resources (Eisenhardt and Martin 2000) would be what management innovation relies on. Since accumulating valuable resources is not enough to support a sustainable competitive advantage in the ever-changing competitive environment (Teece et al. 1997; Teece 2007; Liao et al. 2009), dynamic capability, a capability that firms may possess by virtue of their people or material resources (Teece et al. 1997; Eisenhardt and Martin 2000; Zollo and Winter 2002), may enable firms to reconfigure internal and external competencies to support the project of management changes to address the challenges faced in rapidly changing environments. Therefore, based on the process of adoptive management innovation, this research further explores how dynamic capability of a firm drives adoptive management innovation in Chap. 5. Before doing this, it attempts to address how adoptive management innovation could be measured as a complex and multidimensional concept or process whose results are intangible, uncertain, lagging and even inseparable from that of technological innovation. Since management innovation represents an overall, huge and complex project with the difficulty in output measuring, a process-oriented method based on a sequential-phase process would be more appropriate than a result-oriented one based on ambiguous outputs of innovation. Then, a scale for measuring based on this process-oriented method would be developed through an extraction of characteristics and major activities in each phase of innovation and the reliability and validity assessment. Research literatures emphasize the relationship between dynamic capability and innovation, and contend that a firm's dynamic capability could significantly enhance its ability to innovate (O'Connor 2008), especially in the case of radical management innovation. That is, management innovation is not an isolated project, but relies on dynamic capability of a firm. Though the positive effects of dynamic capability on innovation have been extensively identified, literatures still contain gaps in discussing how dynamic capability internally enhances performance of management innovation. A close look at the body of literature also reveals that a majority of researches are theoretical and conceptual. In consistence with these studies which suggest that dynamic capability is helpful in enacting change in a systematic and fruitful way, this research attempts to explore how different components of dynamic capability affect each phase of adoptive management innovation.

Finally, how intangible (adoptive) management innovation with ambiguous outcomes would affect performance of organizations? Most studies on management innovation advocate a positive effect of management innovation on organizational performance and see it as one of the most important and sustainable sources of competitive advantage for firms (Hamel 2006; Mol and Birkinshaw 2009). However, little research has been done to explore how management innovation would contribute to organizational performance. Considering that management innovation is typically tacit in nature and full of uncertainty and ambiguity, this research argues that, it may contribute to organizations by directly serving tangible product innovation, especially complex products with a separate brand. Researchers have found the value of the alignment between product innovation and management innovation, and argue that this alignment has become a precondition for pursuing maximal performance of organizations (e.g. Daft 1978; Xu and Xie 2004). This indicates that tangible product innovation could produce direct effect on organizational performance, while management innovation may yield more continuous but lagging outcomes. However, what is still puzzling is that how the alignment effect could be produced. Especially for those complex-product firms, how intangible management innovation with ambiguous outcomes could efficiently support the process of product innovation is still a Black-box. This research attempts to identify major management innovation conducted in the innovation process of complex product with a separate brand and explores how these kinds of management innovation support different phases of product innovation. To do so, a deep investigation into the whole innovation process of Lexus of Toyota Motor was conducted, aiming to find out all management innovation activities and analyze their effects. Based on this, it further explores how management innovation supports each phase of product innovation to produce alignment effect.

1.3 Research Structure

This book is organized as following.

This chapter presents a brief introduction on why this research focuses on management innovation especially adoptive management innovation in the developing countries like China where abundant firms depend on adoption of advanced management practices, processes, structures, or techniques to survive and develop, what contributions this research made to both management theories and management practices, how this research was organized and also through what process the research was completed.

Chapter 2 describes the conception of what makes adoptive management innovation unique and a review of literatures on management innovation. Four key perspectives in the literature of management innovation have been identified, including condition perspective, effectiveness perspective, diffusion perspective and process perspective. Condition perspective focuses on what conditions give rise to the emergence and diffusion of management innovation referring to institution, culture, rationality, and human resources; Proponents of effectiveness perspective attempt to explore how to improve performance of management innovation; Diffusion perspective focuses on how new management ideas or methods are delivered; While process perspective focuses on sequential phases through which management innovation occurs. It is worth noting that the research on management innovation based on Chinese enterprises' practices is more systematic and in-depth than that abroad, and the research results are abundant. Perhaps the research hasn't distinguished the adoptive type and all-new type of management innovation, or it has paid more attention and emphasis on the all-new innovation concerning about new knowledge creation, all these studies have laid an important foundation for the exploration of adoptive management innovation.

Chapter 3 focuses on the process through which an adoptive management innovation occurs. Based on the conception of what makes adoptive management innovation unique, this section attempts to investigate into the case of the Organizational Efficiency Management of Jiangxi Mobile in China to identify key activities of adoptive management innovation and develop a two-interlinked-subprocess framework of adoption decision and implementation. The subprocess of adoption decision mainly consists of themes of problem identification, innovation perception, attitude formation, problem diagnoses, innovation revisions, proposal evaluation and selection. The implementation process contains three phases (that is, implementation preparation phase, overall implementation phase and the solidification phase). The results indicate that adoption of existing management practices or methods from somewhere else is a more complex and logical process rather than a simple one of knowledge transferring. It needs to integrate existing practices into new organizational context and establish their innovative values during implementation. One core element of the process framework in this research is the emphasis on activities of problem diagnoses and realization of the fitness between management practices adopted and the new organizational context, and another one is the sequence of activities in the whole process.

Chapter 4 focuses on addressing why some managers pursue innovative opportunities by introducing new management practices while others do not in China, and what characteristics of managers affect their decision of introducing new practices and how. According to some existing research on introduction or adoption process of management innovation, this section sets up a framework of adoption decision and the undertakers, including innovation intention, knowledge acquirement, problem identification, proposal evaluation and yes-no selection, which mainly focuses on the roles of managers in early introduction decision process. Based on an exploration on the decision process of adoptive management innovation and responsibility of managers in each stage, this section abstracts three major activities of managers referring to innovation intention, knowledge acquirement and risk perception, and then establishes a three-dimension decision figure made up by paths from entrepreneurial orientation to innovation intention, social network to knowledge acquirement and cognitive biases to risk perception by further combining with relevant literatures and decision practices. Data on 237 managers from different firms were collected to further examine the model by adopting SEM and the affecting paths, cross functions and the differences of affecting degree among variables were pointed out, with the purpose of showing affecting paths and internal mechanism of complex decision making.

Chapter 5 investigates the process through which organizational routines evolve in implementing management innovation, with existing routines overturned and new routines created and solidified. This chapter makes a deep case analysis of the implementation process of DD innovation in HOAU, reveals the complex mechanism of the implementation from the perspective of routine evolution and establishes the implementation process model of organizational innovation. The results of the case study can be stated in two aspects. Theoretically, it breaks the focus on multidimensional factors or inductive general regularity of an innovation, explores the implementation mechanism of innovation at the most microscopic level; the evolution process of organizational routines is revealed, namely, recreation of new routines is the key for routine evolution, thus for success of management innovation. From the practical view, it confirms management innovation as a complex project concerning organizational routines which represent a central and fundamental element of organizations. Also, it finds that organizational routines evolve in innovation implementation through a three-phase process consisting of the existing-routine-domination phase, the new-routine-creation phase and the new-routine-solidification phase, each exhibiting different innovation activities and characteristics of participants' cognition and behaviors.

Chapter 6 explores the driver of adoptive management from a dynamic capability perspective. Seeing management innovation as one of the most important and sustainable sources of competitive advantage for firms, this section attempts to address how management innovation with intangible, uncertain, lagging and inseparable outputs could be measured, and how dynamic capability of firms internally drives management innovation. Focusing on adoptive management innovation, it resets up a four-phase process consisting of initiation, outside search, proposal establishment and implementation based on the process model in Chap. 3, and develops a new scale for measuring by adopting a process-oriented method. Particularly, building on previous research of how and why management innovation occurs, it focuses on effects of dynamic capability's four components (i.e., sensing capability, absorptive capacity, relational capability and integrative capability) on each phase of innovation. The results of SEM on data from 264 firms in China and analysis through PLS-SEM indicate that, each component affects each phase of the adoptive management innovation process with a unique path from initiation through to implementation, and though the four dimensions of dynamic capability focus on different aspects, they influence organizational performance through interwoven interaction. Actually, in the four dimensions, relational capability based on relationships with firms and individuals serves as a fundamental role and facilitates sensing capability, absorptive capacity, and integrative capability, which reflects the culture in China where affection rooting in relationships determines acquisition of resources and even survival and development of organizations. Besides, what is worth mentioning is that, as innovation results can be intangible and lagging, and consequently difficult to measure, the method offers managers an alternative by monitoring the effects of different capabilities at each stage of the adoptive management innovation process.

Chapter 7 focuses on the effect of adoptive management innovation by addressing what major management innovation may be conducted by a firm when building up a new brand of complex product like automobiles, and how these kinds of management innovation effectively support each phase of product innovation or even the whole process from design and development, to production and to commercialization, aiming to offer an effective path for producing effects of management innovation on organizations. By adopting a longitudinal interpretive and exploratory case study on Lexus of Toyota Motor, a typical complex product with a single brand, it develops a general framework that explicates support and promotion mechanism of management innovation to the process of complex product innovation. Based on existing literatures on management innovation, process of product innovation and the alignment between the two, five explicit management innovation and five implicit management innovation were identifies in the case of Lexus, and a framework showing how these explicit management innovation support the product innovation process through implicit management innovation was set up. That is to say, this work identifies major management innovation when operating a new complex product brand, divides them into two group of explicit (observable management practices) and implicit (unobservable management ideas) management innovation. Accordingly, the results indicate that both explicit management innovation (i.e., observable management practices) and implicit management innovation (i.e., unobservable management concepts) may interdependently coexist in the firm when creating a complex product with a separate brand, and the former may support the process of product innovation through the latter. More specifically, the three major phases of complex-separate-brand-product innovation, each needs the support from different management innovation. Moreover, the results indicate that there exist the synergistic effects of product innovation and management innovation, especially the support and promotion to the whole process of product innovation from management innovation. Finally, Chap. 8 presents a brief conclusion on six major findings obtained from this research, concludes various implications for both theories by addressing the four questions mentioned in the introduction and practice on the process, adoption decision, driver and effects of adoptive management innovation, and lists several limitations and determines future directions.

1.4 Research Framework

The research framework is shown in Fig. 1.1.

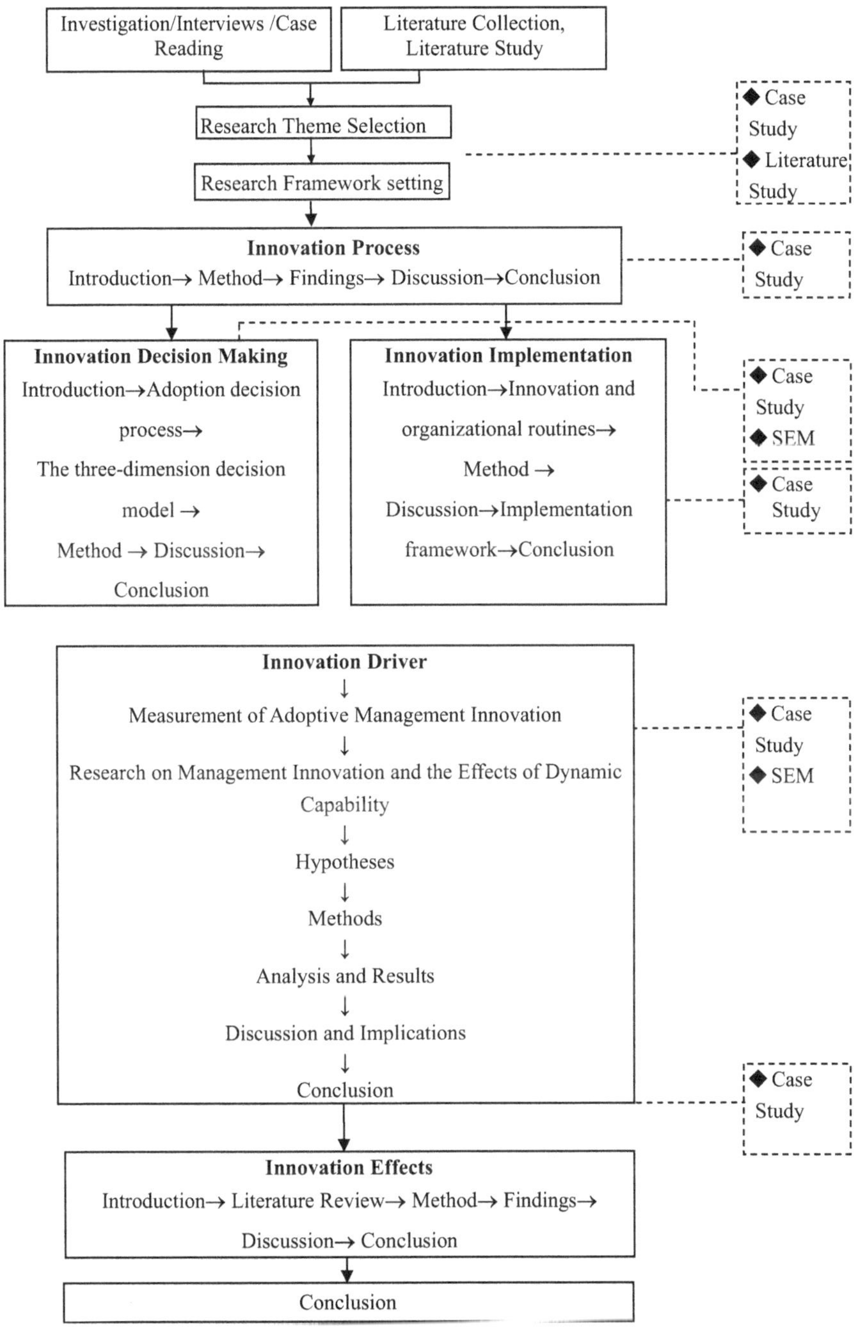

Fig. 1.1 The research framework

References

Abrahamson, E.: Management fashion. Acad. Manag. Rev. **21**(1), 254–285 (1996)
Ansari, S.M., Fiss, P.C., Zajac, E.J.: Made to fit: how practices vary as they diffuse. Acad. Manag. Rev. **35**(1), 67–92 (2010)
Argyris, C., Schön, D.A.: Organizational learning: A Theory of Action Perspective. Birkhäuser Basel (1978)
Barney, J.B.: Firm resource and sustained competitive advantage. J. Manag. **17**(1), 99–120 (1991)
Birkinshaw, J., Mol, M.J.: How management innovation happens. Mit Sloan. Manag. Rev. **47**(4), 81–88 (2006)
Birkinshaw, J., Hamel, G., Mol, M.J.: Management innovation. Acad. Manage. Rev. **33**(4) (2008)
Birkinshaw, J., Crainer, S., Mol, M.: Special report on management innovation. Bus. Strategy Rev. **18**(1), 45–73 (2007)
Burns, L.R., Wholey, D.R.: Adoption and abandonment of matrix management programs: effects of organizational characteristics and interorganizational networks. Acad. Manage. J. Acad. Manage. **36**(1), 106–138 (1993)
Daft, R.L.: A dual-core model of organizational innovation. Acad. Manag. J. **21**(2), 193–210 (1978)
Dahmén, E.: Entrepreneurial activity and the development of swedish industry, 1919–1939, vol. 3. Irwin-Dorsey Ltd, Georgetown Ontario (1970)
Damanpour, F., Schneider, M.: Phases of the adoption of innovation in organizations: effects of environment, organization and top managers. Br. J. Manag. **17**(3), 215–236 (2006)
Damanpour, F., Walker, R.M., Avellaneda, C.N.: Combinative effects of innovation types and organizational performance: a longitudinal study of service organizations. J. Manage. Stud. **46**(4), 650–675 (2009)
DiMaggio, P.J., Powell, W.W.: The iron cage revisited: institutional isomorphism and collective rationality in organizational fields. Am. Sociol. Rev. **48**(2), 147–160 (1983)
Eisenhardt, K.M., Martin, J.A.: Dynamic capabilities: what are they? Strateg. Manag. J. **21**(10–11), 1105–1121 (2000)
Ekvall, G., Arvonen, J.: Leadership profiles, situation and effectiveness. Creativity Innovation Manage. **3**(3), 139–161 (1994)
Elenkov, D.S., Judge, W., Wright, P.: strategic leadership and executive innovation influence: an international multi-cluster comparative study. Strateg. Manag. J. **26**(7), 665–682 (2005)
Feldman, M.S., Pentland, B.T.: Reconceptualizing organizational routines as a source of flexibility and change. Adm. Sci. Q. **48**(1), 94–118 (2003)
Guillén, M.F.: Models of management: work, authority, and organization in a comparative perspective. University of Chicago Press, Chicago (1994)
Hamel, G.: The why, what, and how of management innovation. Harvard Bus. Rev. **84**(2), 72–84 (2006)
Hamel, G.: The future of management. Harvard Business School Press, Boston (2007)
Henderson, R.M., Clark, K.B.: Architectural innovation: the reconfiguration of existing product technologies and the failure of established firms. Adm. Sci. Q. **35**(1), 9–30 (1990)
Howard-Grenville, J.A.: The persistence of flexible organizational routines: the role of agency and organizational context. Organ. Sci. **16**(6), 618–636 (2005)
Howell, J.M., Higgins, C.A.: Champions of technological innovation. Adm. Sci. Q. **35**(2), 317–341 (1990)
Hughes, R.J.M.: The Design and Implementation of Programming Languages. Ph.D. thesis, Oxford University (1983)
Kimberly, J.R., Evanisko, M.J.: Organizational innovation: the influence of individual, organizational, and contextual factors on hospital adoption of technological and administrative innovations. Acad. Manage. J. Acad. Manage. **24**(4), 689–713 (1981)
Kossek, E.E.: Human Resources Management Innovation. Hum. Resour. Manage. **26**(1), 71–92 (1987)

Lazonick, W., Teece, D.J.: Introduction: Management innovation-essays in the spirit of alfred D. chandler, jr. Ind. Corp. Change **19**(2), 295–296 (2010)
Liao, J., Kickul, J.R., Ma, H.: Organizational dynamic capability and innovation: an empirical examination of internet firms. J. Small Bus. Manage. **47**(3), 263–286 (2009)
Lin, H.F., Su, J.Q.: Research on the introducing mechanism of management innovations based on the perspective of internal promoters. Sci. Res. Manage. **31**(6), 136–143 (2010)
Lin, H.F., Su, J.Q.: A case study on adoptive management innovation in China. J. Organ. Change Manage. **27**(1), 83–114 (2014)
Mccabe, D.: 'Waiting for dead men's shoes': towards a cultural understanding of management innovation. Hum. Relat. **55**(5), 505–536 (2002)
Mol, M.J., Birkinshaw, J.: The sources of management innovation: when firms introduce new management practices. J. Bus. Res. **62**(12), 1269–1280 (2009)
O'Connor, G.C.: Major Innovation as a dynamic capability: a systems approach. J. Prod. Innov. Manag. **25**(4), 313–330 (2008)
Parmigiani, A., Howardgrenville, J.: Routines revisited: exploring the capabilities and practice perspectives. Acad. Manag. Ann. **5**(1), 413–453 (2011)
Schumpeter, J.A.: Capitalism, Socialism, and Democracy. Harper and Row, New York (1950)
Stata, R.: Organizational learning-the key to management innovation. Mit Sloan Manage. Rev. **63**(1), 63–73 (1989)
Su, J.Q., Lin, H.F.: Research on the process mechanism of introduced management innovation. Sci. Sci. Manage. S&T **31**(1), 34–41 (2010)
Teece, D.J.: The diffusion of an administrative innovation. Manage. Sci. **26**(5), 464–470 (1980)
Teece, D.J.: Explicating dynamic capabilities: the nature and microfoundations of (sustainable) enterprise performance. Strateg. Manag. J. **28**(13), 1319–1350 (2007)
Teece, D.J., Pisano, G., Shuen, A.: Dynamic capabilities and strategic management. Strateg. Manag. J. **18**(7), 509–533 (1997)
Tushman, M.L., O'Reilly, C.A.: Winning Through Innovation. HBS Press, Boston (2002)
Utterback, J.M.: Mastering The Dynamics of Innovation. Harvard Business School Press, Boston (1994)
Vaccaro, I.G., Jansen, J.J.P., Bosch, F.A.J.V.D., Volberda, H.W.: Management innovation and leadership: the moderating role of organizational size. J. Manage. Stud. **49**(1), 28–51 (2012)
Wernerfelt, B.: A resource-based view of the firm. Strateg. Manag. J. **5**(2), 171–180 (1984)
Xu, Q.R., Xie, Z.S.: study of firm's innovation synergy and the evolvement model. Stud. Sci. Sci. **22**, 327–332 (2004)
Zollo, M., Winter, S.G.: Deliberate learning and the evolution of dynamic capabilities. Organ. Sci. **13**(3), 339–351 (2002)

Chapter 2
Literature Review

2.1 Definition of Adoptive Management Innovation

Hamel (2006, p. 75) identifies management innovation as "a marked departure from traditional management principles, processes, and practices or a departure from customary organizational forms that significantly alters the way the work of management is performed". Birkinshaw et al. (2008, p. 829) further clarify the nature of management innovation by defining it as the "generation and implementation of a management practice, process, structure or technique that is new to the state of the art and is intended to further organizational goals". A little differently, Mol and Birkinshaw (2009, p. 1269) define management innovation as "the introduction of management practices new to the firm and intended to enhance firm performance". Adoptive management innovation represents new approaches to structuring a firm, new management techniques, and new marketing methods that firms adopt from another context (Mol and Birkinshaw 2009). According to these definitions and existing literatures on management innovation, when seek to develop an operational definition of management innovation we should focus on three core issues: what exactly is being innovated? How new does an innovation have to be? And what is the purpose of management innovation?

The first issue is about the content of innovation. Birkinshaw et al. (2008) advocate a separation of two analysis levels: at the more abstract level are management ideas, defined by Kramer as "fairly stable bodies of knowledge about what managers ought to do… a system of assumptions, accepted principles and rules of procedure" (1975, p. 47); while at a more operational level are management practices, management processes, management techniques, and organizational structures, namely different facets of the rules and routines by which work gets done inside organizations. In line with Birkinshaw et al. (2008), abundant of research focus on observable innovation in management practices, processes and structures at the operational level rather than management ideas or ideologies, "because this is the level at which observable changes take place in the way work is done and the

H. Lin, *Adoptive Management Innovation*,
https://doi.org/10.1007/978-981-10-7712-8_2

management innovation process can be witnessed" (Birkinshaw et al. 2008, p. 828). However, management ideas (e.g. Total Quality Management, Organizational Learning, Organizational Culture), though unobservable, "can serve as cognitive tools that managers use to sort out the complexities of reality, frame the relevant issues, and choose among alternative paths of action" (Guillén 1994, p. 4), which indicates a fundamental effect on performance of firms (e.g. Guillén 1994; Barley and Kunda 1992; Abrahamson 1996; Suddaby and Greenwood 2005). When exploring the process and driver of management innovation, this research prefers to observable innovation in management practices, processes and structures at an operational level. However, when exploring how management innovation supports the process of product innovation to produce effect, it focuses on both management practices and ideas by conquering the unobservable problem through a deep investigation into practices of the sample case enterprise.

The second issue is about how new an innovation has to be. Some researchers (e.g. Abrahamson 1996; Kimberly and Evanisko 1981; Birkinshaw et al. 2008) pay more attention to new-to-the-state-of-the-art innovation without known precedent. For the primary reason, this is the area where existing knowledge is the most limited. Consequently, Birkinshaw et al. (2008, p. 829) define management innovation as "the generation and implementation of a management practice, process, structure, or technique that is new to the state of the art and is intended to further organizational goals", which attracts the attention of a large number of theorists and industrialists. However, only a few enterprises adopt generative management innovation because of its uncertainty of the outcome. Therefore, in the following research, some others (e.g. Vaccaro et al. 2012; Mol and Birkinshaw 2009) are attracted by the new-to-the-organization innovation, namely adoptive management innovation that has been successfully implemented somewhere else. Actually, adoptive management innovation represents new approaches to structuring the firm, new management techniques and new marketing methods that firms pick up from another context (Mol and Birkinshaw 2009). This is consistent with the argument of Angle and Ven (2000) that the adoption of innovation basically means that the innovation is novel to the adopting unit intending to derive anticipated benefits from changes that the innovation may bring to the organization (West and Anderson 1996), or the adoption of an idea or behavior that is new to the adopting organization (Damanpour and Schneider 2006). In other words, adoptive management innovation is more common and has a stronger practical significance.

The third issue is about the purpose of management innovation. Some researchers regard management innovation as having little lasting impact on the organization (e.g. Abrahamson 1996), whereas most of them view management innovation as generating positive outcomes for the innovating firm and/or for society as a whole (e.g. Birkinshaw et al. 2008; Mol and Birkinshaw 2009). This research views management innovation as intending to further the organization's goals, which may include both traditional aspects of performance (e.g. financial goals) and softer aspects (e.g. employee satisfaction). That is, the purpose of management innovation is to produce positive effect on organizational goals and performance. This is appropriate because it helps to explain why firms are prepared to engage in the costly and somewhat risky process of management innovation in the first place. This approach serves to underline the important point that not all management innovations are ultimately successful. Moreover, it should also be noted that goals are rarely entirely exogenous to the organization; indeed, the process of innovating can result in the introduction of new practices or programs.

Following Vaccaro et al. (2012) and Mol and Birkinshaw (2009), this research focuses on adoptive management innovation for three reasons. First, according to Vaccaro et al. (2012), as for innovations new to the state of the art or new to the world, the level of analysis is management at large, and in the case of "new to the organization", the level of analysis is the firm, focusing on which level of analysis enables us to empirically test a series of hypotheses at the firm level of analysis and draws on a potentially much more sizable sample of management innovation. This research intends to go deep into the micro world of organizations by focusing on innovation activities on the firm level. Second, it is adoptive management innovation that dominates the innovation practices of Chinese firms and even firms around the world, for example, the extensive introduction and implementation of Total Quality Management (TQM), Business Process Reengineering, Strategic Change, Customer Relationship Management programs, which might be categorized as a management innovation (e.g. McCabe 2002; Zbaracki 1998). Third, the introduction and implementation of management innovation that is novel to an organization is also of high uncertainty. The success of new practices may rely on their adaptation to their idiosyncratic context within the organization where they are adopted (Ansari et al. 2010; Vaccaro et al. 2012), which indicates a high value in focusing on this body of innovations. Therefore, research on the adoptive management innovation may offer invaluable insights.

In comparison with generative management innovation proposed by Birkinshaw et al. (2008) and considering the nature of adoptive innovation, this research defines adoptive management innovation as *the introduction and implementation of an existing or mature management practice, process, structure, or technique that has been not only implemented somewhere else successfully but also intended to improve operation efficiency and organizational performance and further organizational goals.*

Moreover, considering that both adoptive and generative management innovation affect product innovation and organization operation in a similar way, the author concerns both when exploring their effects in Chap. 6.

2.2 Why Management Innovation Matters

Considering the relationship between management innovation and organizational performance, different scholars have different opinions. Although a few scholars insist that the effect of management innovation is negative (McCabe and Knights 2002; Staw and Epstein 2000) and even consider it as a way to control employees or a fashion-driven process by which consultants seek benefits by promoting new practices (Abrahamson 1996), most scholars support the perspective that management innovation has a positive impact on organizational performance. Generally, they believe that management innovation is an important way to improve competitiveness (Caroli and Reenen 2001; Piva and Vivarelli 2004). There are two specific reasons why they believe so: on the one hand, the success of technical product innovation and technological innovation to a large extent depend on the adjustment and adaptation process innovation of organization structures and internal operating procedures so that management innovation is a prerequisite and booster to effectively implement technology products and technological innovation; on the other hand, management innovation has an important impact on productivity, delivery time, quality and flexibility, and therefore represents the direct source of organizational competitive advantages. After a series of intentional actions, management innovation is implemented, becomes an integral part of the organization's capabilities (Edmondson 2003), and gradually evolves into organization routines. Most studies suggest that management innovation is linearly dependent on organizational performance (Edmondson et al. 2001). Indeed, management innovation has a positive impact on the organization through changing its daily behaviors or actions to improve its efficiency, quality and productivity (Staw and Epstein 2000). And the greater the implementation intensity of management innovation is, the greater the increase of organizational performance is (Edmondson et al. 2001). However, a small number of studies have opposing views. Take the study of Naveh et al. (2006) for example, they believe that there is an inverted U-shaped relationship between management innovation level and organizational performance. According to the perspective of organizational performance, inadequate or excessive implementation of management innovation can both aggravate the organization performance, but management innovation can actively promote the performance of the organization as long as choosing appropriate new management practices and investment degree.

In addition to the organizational performance, according to Birkinshaw and Mol (2006), management innovation is able to contribute to a firm's long-lasting competitive advantage. They took the history of competition and leadership in the car industry for example. At first glance, one might expect the long-run changes in industry leadership to be driven by technological changes or the introduction of breakthrough products. But in reality, these factors have played a relatively minor role. Toyota, the undisputed current industry leader, grew into a position of dominance largely on the basis of its lean manufacturing platform, and through specific management innovation such as Kanban, Target Costing and Just-In-Time. Although Toyota cars were technologically advanced, that is not what has distinguished the company from its competitors. Before Toyota, the global industry leader was General Motors, which had achieved dominance in large part through its invention of the divisionalised "M-form". That allowed GM to achieve unprecedented levels of growth (and allowed the divisions to segment their markets according to different brands and models—another important management innovation). These innovations, in turn, helped GM overtake Ford Motor Company, which famously stuck to its Model T until the bitter end, despite customers' appetite for greater choices. Henry Ford had, of course, steered his company to fame and initial industry leadership through the invention of the moving assembly line, which streamlined the management of production operation. In the automobile industry, then, it was management innovation, not technological innovation, which drove long-term competitive advantages.

In fact, since Ford proved the value of the Moving Assembly Line in increasing the productivity of its workers, GM showed the value of Market Segmentation in encouraging the creation of products that were closer to actual customer demand, or Toyota proved its Just-In-Time System efficient in improving production, these management practices or methods have been adopted and implemented by abundant firms around the world. Although Birkinshaw and Mol (2006) state that the M-form organization of GM took other companies decades to implement for two reasons that the M-form organization suited GM's needs particularly well at the time and M-form required lots of co-ordination processes between headquarters and the divisions that were hard to copy, in fact, lots of firms around the world have benefited from this new management practice. To be noticed, as the definition of adoptive management innovation indicates, to adopt new practices or methods from somewhere else is not a easy copy process, but a more complicate one of combining them with characteristics of a certain firm.

In China, more and more firms survive and stand out in competition, such as Haier, Legend, China Mobile, Galanz, Alibaba and Baidu. Then, what makes these Chinese firms success? Great products? Yes. Great people? Sure. Great leaders? Usually. But if you dig deeper, there is a more fundamental reason: management innovation, especially adoptive management innovation.

Haier Global (Haier). Haier was founded in 1984 in Qingdao, China. Through its entrepreneurial and innovative spirits, Haier has transformed itself from an insolvent collectively-owned factory on the brink of bankruptcy into the number one global home appliance brand in the last 30 years. Haier is known for disruptive innovation in its product solutions and management models, for example, the OEC management control system, unique performance management systems, the market-chain-based business process reengineering system and Win-win Model of Individual-Goal Combination. In particular, Haier's restructuring effort on the market-chain-based business process reengineering system which began in late 1998 has taken it from a nearly bankrupt factory to a company with global revenue of RMB180.3 billion (USD 29.5 billion) in 2013. Haier began to implement the market-chain-based business process reengineering system in late 1998. According to CEO Zhang Ruimin, a market chain is a series of business process activities to make products or render services to satisfy customers' needs. In a nutshell, a market chain links every employee's work with the market, which can be an external or internal market. Therefore, every Haier employee's next downstream activity or process is a market, and every employee faces a market with a direct link to a customer. This allows the firm to convert external market competition into a type of internal competition. Therefore, with employee compensation tied to market performance, every employee provides the best performance to meet his or her customers' needs. To do so, every Haier employee has a picture of the entire organization that shows how company parts interrelate. For example, the production department's direct customer is the distribution department. If you ask an employee where an order comes from, he or she can tell you. To understand the company's entire market-chain system, each employee attends training at Haier University and learns everything from product development to production and distribution.

Lenovo Group (Lenovo). Lenovo Group, which was invested 200,000 Yuan RMB by the Computing Institute of Chinese Academy of Science and founded by 11 scientists and technicians in 1984, now has developed into a multiple large-scale enterprise group in IT industry. Now the head office of Lenovo is located in Purchase, New York, and two major operating centers in Beijing, China and Raleigh of North Carolina, the USA. The sales network of the new Lenovo has spread all over the world through the sales organization of its own, its business partners, as well as the alliance with IBM. Lenovo has about 19,000 staff throughout the globe. Its research and development centers are scattered in Beijing, Shenzhen, Xiamen, Chengdu and Shanghai, China, and Tokyo, Japan and Raleigh of North Carolina, USA. As a leading enterprise of PC market in the world, Lenovo is engaged in developing, producing and selling the most reliable, secure and easy-to-use technical products, superior and professional services and helping the users and partners throughout the world to succeed. Moreover, the development of Lenovo relies on series of new management practices. Lenovo has dedicated mergers and acquisitions team that tracks the progress of these integrations. Lenovo has an annual meeting where the management of newly acquired companies meets with its top 100 executives. In these meetings, where English is the medium, Lenovo explains its global strategy and how new executives fit in. For example, Lenovo acquired IBM's personal computer business in 2005, including the ThinkPad laptop and tablet lines. Lenovo's acquisition of IBM's personal computer division accelerated access to foreign markets while improving both its branding and technology. Lenovo paid US$1.25 billion for IBM's computer business and assumed an additional US$500 million of IBM's debt. This acquisition made Lenovo the third-largest computer maker worldwide by volume. In June 2011, Lenovo announced that it planned to acquire control of Medion, a German electronics manufacturing company. The acquisition doubled its share of the German computer market, making it the third-largest vendor by sales (after Acer and Hewlett-Packard). The deal, which closed in the third quarter of the same year, was the first in which a Chinese company acquired a well-known German company.

China Mobile Group (China Mobile). As the leading mobile services provider in Mainland China, the Group boasts the world's largest mobile network and the world's largest mobile customer base. In 2013, the Company was once again selected as one of the "FT Global 500" by Financial Times and "The World's 2000 Biggest Public Companies" by Forbes magazine, and recognized on the Dow Jones Sustainability Emerging Markets Index. The Company currently has a corporate credit rating of Aa3/Outlook Stable from Moody's Investor Service and AA-/ Outlook Stable from Standard & Poor's, equivalent to China's sovereign credit rating, respectively. It engaged in offering voice services (e.g., local calls, domestic long distance calls, international long distance calls, intra-provincial roaming, inter-provincial roaming, international roaming and voice value-added services), data services (for example, SMS and MMS, Wireless Data Traffic, Applications and Information Services) and the customer service (i.e., nationwide free customer service hotline "10086") like most of other mobile services providers. It has gone through a rapid-developing period, which could be attributed to the efforts made to introduce and implement various management practices or methods in response to environmental changes in its competitive market. For example, the Company's wholly-owned subsidiary, CMC, entered into a share subscription agreement with ANHUI USTC IFLYTEK Co. Ltd ("Anhui USTC") to acquire 70,273,935 ordinary shares of Anhui USTC, representing 15% of its enlarged issued share capital, in an effort to speed up our mobile Internet deployment; The Company launched the new commercial brand "and!" which represents China Mobile's belief in continuously seeking value innovation, promoting industry developments and accomplishing the strategic vision of "Mobile Changes Life" in 2013; China Mobile (Hong Kong) Limited completed the acquisition and privatization of former China Resources Peoples Telephone Company Limited and later changed its name to China Mobile Peoples Telephone Company Limited in 2006.

Alibaba Group Holding Limited (Alibaba). Alibaba is a Chinese e-commerce company that provides consumer-to-consumer, business-to-consumer and business-to-business sales services via web portals. It also provides electronic payment services, a shopping search engine and data-centric cloud computing services. The group began in 1999 when Jack Ma founded the website Alibaba.com, a business-to-business portal to connect Chinese manufacturers with overseas buyers. In 2012, two of Alibaba's portals handled 1.1 trillion yuan ($170 billion) in sales. Alibaba's consumer-to-consumer portal Taobao, similar to eBay.com, features nearly a billion products and is one of the 20 most-visited websites globally. The Group's websites accounted for over 60% of the parcels delivered in China by March 2013, and 80% of the nation's online sales by September 2014. Alipay, an online payment escrow service, accounts for roughly half of all online payment transactions within China. Alibaba reported sale of more than $9 billion on China's Singles' Day in 2014.

In fact, more and more examples of adoptive management innovation emerge all over China. For example, The National Enterprise Management Modernization Innovation Achievements annually awarded by China Enterprise Association Management Modernization Work Committee have listed out abundant emerging management innovation cases since 1991, most of which are adoptive management innovation implemented by firms from different industries. Table 2.1 shows some of the innovation examples from the National Enterprise Management Modernization Innovation Achievements in 2014.

Besides innovation achievements selected and awarded by the national and local governments of China, abundant other innovations have been set up as examples. For example, PKU Business Review, a business journal sponsored by the Management Case Study Center of Guanghua School of Management, Peking University, and Peking University Publishing House, set up China Academy of

Table 2.1 Some examples of adoptive management innovation from the national enterprise management modernization innovation achievements in 2014

Firms	Management innovation
China Aerospace Science and Technology Corporation	Integrated engineering management
Nanjing Kangni Mechanical & Electrical Co., Ltd.	Intellectual property strategy management
Dayawan Nuclear Power Operation and Management Co., Ltd.	Excellent operation management
China National Offshore Oil Corporation	Cross-border merger and acquisition
Wahaha Group Co. Ltd.	Quality chain management
China National Machinery Industry Corporation, Ltd.	Business and resource integration management
Shenhua Group Corporation Limited	Financial management
Beixin Group Building Materials Limited by Share Ltd.	Strategic management
Changan Chongqing Automobile Limited by Share Ltd.	Quality management
Industrial Commercial Bank of China Ltd.	Customer precision marketing management
Petro China International Exploration & Development Company	Risk management
Shenyang Aircraft Industry (Group) Co., Ltd.	Agility management
Zhejiang Jieliya Limited by Share Ltd.	Multi brand building and management
China Huaneng Group Limited Hong Kong Company	Cultural management
Guangdong Power Grid Corporation	Performance management
Changchun Railway Passenger Car Limited by Share Ltd.	Business process reengineering
Hubei Sanhuan Forging Equipment Co., Ltd.	R&D management
China Yangtze Power Co., Ltd.	Production management
China Mobile Communication Group Tianjin Co., Ltd.	Service management
China Petroleum Chemical Co of Shanghai Gaoqiao Branch	Knowledge management
China Academy of Launch Vehicle Technology	Production and operation management
Shanghai Railroad Bureau	Integration management
Hebei Aviation Investment Group Co., Ltd.	Integrative service management
The third oil production plant of Changqing Oilfield Company Petro China Co., Ltd.	Lean management
Changhe Aircraft Industry (Group) Co., Ltd.	Information management
Jiangsu Yangnong Chemical Limited by Share Ltd.	Social responsibility management
Zhejiang Tobacco Industry Co., Ltd.	Production digital management

Management Awarding 2006, with the purpose of selecting innovation examples from management practices of firms and extending them all over China. After then, more than twenty firms and their innovations are awarded every year through a complicated evaluation process. Different from those awarded by the government, most China-Academy-of-Management-Award winners are private enterprises, rather that state-owned ones. For example, the fourth awards in 2010 went to thirty excellent firms such as China Minsheng Banking Corp. Ltd., Suning Yun Group Limited by Share Ltd., and Mengniu Dairy (Group) Limited by Share Ltd., for their contributions to localization management. Specific innovations implemented by these firms are shown in Table 2.2.

Table 2.2 The fourth China Academy of Management awards

Firms	Management innovation
China Minsheng Banking Corp. Ltd.	Minsheng division model
Suning Yun Group Limited by Share Ltd.	Business chain development model
Mengniu Dairy (Group) Limited by Share Ltd.	Fine management and brand establishment
Sany Group	Market competition strategy
Citic Trust	Non boundary service management model
CPMC Holdings Ltd.	Integrated service marketing model
Zhuoda Group	Multi-value real estate model
Amway China	Store operation ladder type training system
Inner Mongolia First Machinery Group Corporation	Industrial chain optimization and industrial clusters construction
China Aerospace Science & Industry Co., Ltd.	Safe operation model
Wantong Real Estate	Value management
China National Automotive Industry International Corporation	Asset restructuring
China Oilfield Services Ltd.	Scenario planning system construction
Sunrain Group Corporation	Cross-border integrated marketing model
Fudian Bank	Cross-border business development
Xinxing Ductile Iron Pipes Co., Ltd.	Integration of management and control
Fotile Group	Strategic management
Green Group	Employee keeping project
Busen Group	Sound development management
Guangzhou Echom Science & Technology Co., Ltd.	Design and manufacture services innovation industry model
Tuopai Group	Ecological management model
Tader Coal SCM Ltd.	Integrated management model
Zhejiang Daily Group	Brand strategy management
Aigo Digital Technology Corporation	Cooperation management model
True Kung Fu Restaurant Management Co., Ltd.	Fast food standardization management
EVOC Intelligent Technology Co., Ltd.	Non classical management model

(continued)

Table 2.2 (continued)

Firms	Management innovation
Lanxum Technology Co., Ltd.	Outsourcing business model
Raflatac Enterprise Management Consulting Co., Ltd.	Cultural management
Sinar Mas Paper (China) Investment Co., Ltd.	Management by olympic system
NVC Lighting Technology Co. Ltd.	Channel innovation management

Moreover, within some industries, in order to improve the management level of the whole industry, organizations like the industry federation also set up management innovation awards. For example, in the electric-power industry in China, the industry federation set up a committee responsible for evaluating management practices of its member firms and offering awards to these who obtain outstanding performance through implementation of advanced management methods or concepts. In 2015, the first prizes were awarded to eighty firms or branches for their management innovation achievements, such as State Grid for its construction of a strategic management and control system, China Southern Power Grid for its construction of a new business management system, and Datang Group for its innovation in the employment system. Even in some large groups, their branches are encouraged to improve the management level by themselves. For example, China Mobile Group awards its branches every year for their management practices.

As above examples show, a management breakthrough can deliver a potent advantage to the innovating company and produce a seismic shift in industry leadership. Technology and product innovation, by comparison, tend to deliver small-caliber advantages. More specifically, a management innovation creates long-lasting advantages when it meets one or more of three conditions: The innovation is based on a novel principle that challenges management orthodoxy; it is systemic, encompassing a range of processes and methods; and it is part of an ongoing program of invention, where progress compounds over time (Hamel 2006).

2.3 Research Perspectives on Management Innovation

In its broadest sense, management innovation has certainly received considerable research attention over the years (e.g. Birkinshaw et al. 2008; Mol and Birkinshaw 2009; Vaccaro et al. 2012). As discussion in the following, four different key perspectives in the literature have been identified, including condition perspective, effectiveness perspective, diffusion perspective and process perspective. They will be described in details in the following sections.

2.3.1 Condition Perspective

Condition perspective focuses on what conditions give rise to the emergence and diffusion of management innovation referring to institution, culture, rationality, and human resources.

The institution theory emphasizes the social economic conditions in which new management concepts and practices emerge. For example, Guillén (1994) examines the impact of various institutional factors on the introduction of new managerial ideologies and techniques; Kossek (1987) examines industry- and firm-level influences on the emergence of human resource management innovation; Cole (1985) focuses on how the balance among labor market incentives that are mostly set by the state, the relative strength of industry associations, and the predisposition of organized labor influenced the introduction of small-group activities in different countries. Though these studies are not concerned with the roles of human resources in the process of producing management innovation ideologies, they stress the importance of the preconditions in which innovations or the factors for innovation adoption emerge. Studies on cultural condition refer to the effect of organization culture on successful introduction and implementation of management innovation. It operates at the meso level of analysis by looking at how individual attitudes toward management innovation interact with the organization level introduction of the innovation (Birkinshaw et al. 2008). One strand of this literature takes a critical perspective(Knights and McCabe 2000; McCabe 2002) while the other adopts an intra-organizational process perspective (Stjernberg and Philips 1993; Zbaracki 1998), but both share some common themes: are cognition that established organizations do not change easily, that management innovation has both rhetorical and technical components, and that the outcome of the introduction of a management innovation is rarely what was intended by the senior executives who introduced it rather than culture of the whole organization. Studies on cultural condition provide some insight into how management innovation is implemented, though primarily from the point of view of those who are being asked to participate in the process, rather than those who are driving it. The rationality theory builds on the premise that management innovation is adopted by individuals to make their organizations more efficient and effective. That is, an individual puts forward an innovative solution to address a specific problem that the organization is facing, and he or she then champions its implementation and adoption (Burgelman 1983; Howell and Higgins 1990). Or, it is managers that address critical problems by creating or adopting new practices and offering support for implementation (Howell and Higgins 1990). Some studies favor the case study methodology, while others prefer to quantitative study relying on large samples.

Finally, studies on human resources focus on how internal employees affect management innovation. In order to investigate this, Osterman (1994) collects considerable data on employees from different organizations all over America. Moreover, Chi et al. (2007) focus on employees from certain area or industry. Both draw the same conclusions that: employee training is the main driver for adoption and diffusion of management innovation in organizations, especially human resource innovation; further, employees with more autonomies and responsibilities would easily accept innovations. Meanwhile, other studies have investigated the role of labor unions in acceptance of innovations. They argue that labor unions enable employees from these enterprises to accept innovations more easily since they represent an effective channel for employees to complain and make their jobs safe (Lynch 2007).

Additionally, there are other literatures arguing that management innovation is affected by multiple factors, rather than a single factor. For example, building on the organizational reference group literature, Mol and Birkinshaw (2009) show that management innovation is a consequence of a firm's internal context and of the external search for new knowledge. Furthermore they demonstrate a trade-off between context and search, in that there is a negative effect on management innovation associated with their joint occurrence, and management innovation is positively associated with firm performance in the form of subsequent productivity growth. Damanpour and Schneider (2006) examine the effects of environmental, organizational and top managers' characteristics on the initiation, adoption decision and implementation of innovation, and find that while each dimension accounts for unique variance in the adoption of innovation, organizational characteristics and top managers' attitudes toward innovation have a stronger influence than environmental and top managers' demographic characteristics. Waarts et al. (2002) analyze the dynamic factors that influence the introduction of innovation and indicate that the factors will change with the diffusion of innovation, which mainly consist of innovation characteristics, features of decision makers, internal environment and external environment characteristics.

2.3.2 *Effectiveness Perspective*

Proponents of effectiveness perspective attempt to explore how to improve performance of management innovation. Some provide insights into how organizational learning affects management innovation production and performance. Since adopted by Stata in 1989 into the area of management innovation, organization learning has been viewed as an efficient means of necessary behavior changes for improving performance (Garvin 1993) and realizing new management models for enterprises (Barker 1999) (e.g., a bilateral process model set up by Zhang (1999) to describe relationships among learning, knowledge management and organization innovation). Additionally, Li's (2007) research indicates that organization learning level positively affects management innovation performance; and learning approaches affect management innovation differently; also, significant differentiation of organizational structure, culture and environment leads to that of management innovation performance, organization learning levels and learning methods; and organizational structure, culture and environment also influence the relationship between organizational learning style and management innovation effectiveness. In brief, organizational learning is an important way to promote management innovation or improve management innovation performance, which is an important link in management innovation system research. Further, according to March (1991), organizational learning includes two types: one refers to the exploratory learning; the other is the development learning. Researches on the relationship between organizational learning and innovation are mainly oriented towards technological innovation. Continuing the logic of March (1991), Lin and Su (2012) point out that either exploration or development is accompanied by innovation. On that basis, Lin and Su (2012) propose that building a two-element organization that can coordinate two opposing learning types can create organizational situations and enhance the effectiveness of top managers based on organization dual structure. Others attempt to illustrate effects of dynamic capability on management innovation. Dynamic capability, the process to integrate, reconfigure, gain and release resources (Eisenhardt and Martin 2000), or a capability that may not only be possessed by firms through virtue of their people or material resources (Teece et al. 1997; Zahra and George 2002) but also enable firms to reconfigure internal and external competencies to address the challenges faced in rapidly changing environments. Besides marketing and technological changes (Teece 2007), management innovation, a systematical change of the management model by absorbing existing management practices from the external to address rapidly changing environments and alter internal ways for management work, may be a good way to present the value of dynamic capabilities. In another word, dynamic capabilities of organizations may facilitate the progress of introducing and implementing feasible management practices (Teece 2007). Management innovation by itself is insufficient to get success (Teece 1986), but should be supported by dynamic capabilities to purposefully create, extend or modify its resource base. For example, Kohlbacher (2013) empirically examines the impact of dynamic capabilities on innovation performance through continuous improvement; Lazonick and Prencipe (2005) seek to analyze the role of dynamic capabilities in sustaining the entire innovation process, and argue that innovation depends on "strategic control" and "financial commitment" (two dynamic capabilities); Liao et al. (2009) argue that the firm's ability to mobilize its resources and capabilities and align them dynamically with changing opportunities is of vital importance to constantly innovate, survive and create a competitive advantage; Wu et al. (2012) show that dynamic capabilities facilitate firms' strategic changes toward sustainability and higher competitive advantage through searching, prioritizing, positioning, planning, modifying, and leveraging. In summary, researchers emphasize the relationship between dynamic capabilities and management innovation throughout the process, and argue that a firm's dynamic capabilities could significantly enhance its ability to innovate (O'Connor 2008). Chapter 6 of this research will further explore how dynamic capability of a firm affects the whole process of management innovation.

2.3.3 Diffusion Perspective

Diffusion perspective focuses on how new management ideas or methods are delivered. One strand takes an internal-oriented perspective, namely management fashion (Abrahamson 1996); while the other adopts an external-oriented perspective, namely innovation diffusion (Teece 1980).

Studies on management fashion attempt to understand how the interaction between the supplier and the demander leads to the emergence of management innovation (Abrahamson 1996), which provides abundant insights into the formation and diffusion of management innovation. Based on the exploration on the process of management fashion, Abrahamson (1996) define its essence and present that it is the supplier or producer of management fashions that continuously convince followers of the functions of new management methods or techniques in improving management. Therefore, management fashion is the result of this interacting process, for example, a management technique convinced by followers and a transitory collective brief. Consequently, management fashion has been defined as management techniques or innovations. Carson et al. (2000) collect 16 kinds of popular management fashions in the latter half of the 19th century, such as Management by Objective, Total Quality Management and Flatter Structure, all of which are management innovation. The research on management fashion spans the macro and micro level, focusing on not only the supply industry of management innovation but the individual behavior of managers who adopt the management innovation.

In contrast, diffusion means the spreading of innovation among potential organizational users. Most of the current studies on innovation diffusion are concerned with technology innovation exclusively. It can be traced back to the innovation theory of Schumpeter which defined extensive imitation of technology innovation as diffusion. Not until 1980 did Teece first introduce diffusion theory into the management innovation field and argued that both diffusion modes of management innovation and technological innovation are similar. That is to say, early adopters of both technology and management innovation could get more opportunities to make profits while others would face greater challenges (Rogers 1995). Furthermore, as a higher amount of adopters start to increase the speed of the diffusion or the diffusion process grows exponentially, a phenomenon called the bandwagon effect or diffusion one takes affect (Leibenstein 1950). This phenomenon represents the pressure created by the adopters or rejecters of an innovation for others to follow their lead in treating this innovation (Rogers 1995). In details, the higher the amount of people that decide to follow the bandwagon, the more other people in turn will be decoyed (Leibenstein 1950), which leads to a reinforcing process of management innovation diffusion. However, the diffusion paths of management and technology innovation are not exactly the same. For one hand, without the protection of patent, management innovation can be imitated more freely and smoothly, while technology innovation is protected from being imitated. This leads to the lack of investment in management innovation while the adoptive management innovation prevails. For another hand, management innovation needs considerable startup costs, reorganization of enterprises and reallocation of functions and responsibilities, while technology innovation does not. Hence, the diffusion of management innovation may be slower but more random. Management innovation diffusion emphasizes the diffusion process of innovation outcomes across organizations, industries or countries, which is the process of innovation being widely accepted and adopted. Management innovation diffusion belongs to the last link in the process of management innovation. The research perspective has been largely divorced from the micro-level of enterprise management innovation, but is based on industry and even national level.

2.3.4 Process Perspective

Process perspective focuses on sequential phases through which management innovation occurs, which is most close to the first question of this research. Rogers (1995) first defines the process of innovation adoption (especially technological innovation) as the process through which an individual passes from acquiring knowledge of an innovation to forming an attitude toward the innovation, to a decision for adopting or rejecting, to implementation and use of the new idea, and finally to confirmation of this decision. Consequently, the process of innovation adoption in organizations has been usually divided into a variety of similar activities by other authors, which has formed the basis of most empirical studies on innovation process (e.g. Ahire and Ravichandran 2001; Cooper and Zmud 1990; Grover and Goslar 1993; Rogers 1995). Different terms for the phases have been proposed, for example, evaluation, initiation, implementation and routinization (Hage 1980); awareness, selection, adoption, implementation and routinization (Klein and Sorra 1996); knowledge awareness, attitudes formation, decision, initial implementation and sustained implementation (Zaltman et al. 1973); and initiation, development, implementation and termination (Angle and Ven 2000). These activities can be assembled into three more general phases of pre-adoption, adoption decision and post-adoption, often referred to as initiation, adoption (decision) and implementation (Rogers 1995; Pierce and Delbecq 1977; Damanpour and Schneider 2006). Though these studies mainly focus on the adoption process of technological innovation, they offer some valuable insights for research on that of management innovation.

Based on these studies on innovation adoption, Hamel (2006) first investigates the process of management innovation and argues that a systematic process for producing bold management breakthroughs must include commitment to a big management problem, novel principles that illuminate new approaches, a deconstruction of management orthodoxies, and analogies from atypical organizations that redefine what's possible. Following Hamel (2006), Birkinshaw et al. (2007) develop a five-stage process of management innovation: dissatisfaction with status quo, inspiration from outside, change agents, invention and internal validation.

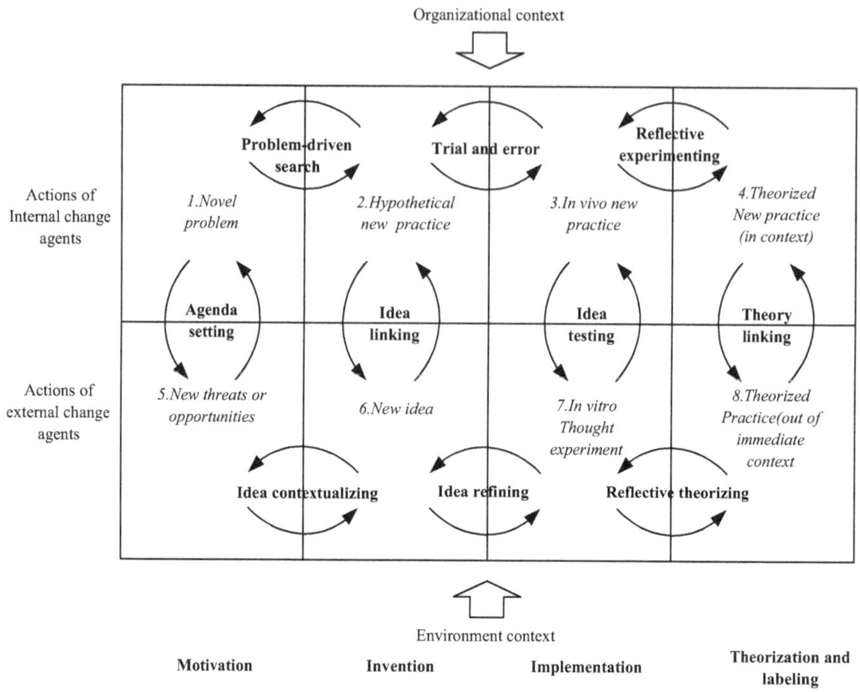

Fig. 2.1 (Generating) Management innovation process framework. *Source* Birkinshaw et al. (2008, p. 832)

They also find that internal validation and external one are two important points of difference that make management innovation a distinct process.

By further investigating a large number of new-to-the-state-of-the-art innovation around the world, for example, Modern Research Lab, Divisional Form, Toyota Production System, Total Quality Management, Discounted Cash Flow and Spaghetti Organization, Birkinshaw et al. (2008) develop a framework with two dimensions that highlights four interlinked phases of the process and the roles played by two key sets of stakeholders. On the one hand, the horizontal dimension consists of four phases of the innovation process: motivation, invention, implementation and theorization and labeling, as showed in Fig. 2.1. More specifically, "Motivation is concerned with the facilitating factors and precipitating circumstances that lead individuals to consider developing their own management innovation; invention reflects an initial act of experimentation out of which a new hypothetical management practice emerges; implementation represents the technical process of establishing the value of the new management innovation in vivo; and theorization and labeling is a social process whereby individuals inside and outside the organization make sense of and validate the management innovation to build its legitimacy" (Birkinshaw et al. 2008, p. 831). On the other hand, the vertical dimension consists of two groups of individuals: internal change agents (i.e., the employees of the innovating company who are proactive in creating interest in, experimenting with, and validating the management innovation in questions) and external change agents (i.e., independent consultants, academics, and gurus who are proactive in creating interest in, influencing the development of, and legitimizing the effectiveness and retention of new management practices) (Birkinshaw et al. 2008).

In addition, this framework illustrates how individuals iterate between the adjacent cells in this frame structure by identifying ten interlinked core activities, and also stresses the important role of context in shaping management innovation. The model not only emphasizes the phases of the innovation process, concluding that management innovation is a process of interlocking, which forms a continuous uninterrupted overall from beginning to end, but also emphasizes the role of internal and external agents and the interaction between them. At the same time, the model also points out that internal agents and external agents interact at each phase of management innovation to facilitate its implementation: In the motivation phase, the two agents interact with the schedule; In the invention phase, they communicate new ideas; The implementation phase shows a circular interaction between practice and theoretical thoughts; In the last phase, theoretical links are carried out on the basis of both internal and external theorization and finally internal and external agents reach accordance.

2.4 Research Issues and Research Methods

As summarized on Table 2.3, four research perspectives on management innovation focus on conditions in which management innovation emerges and internally diffuses, paths for improving the performance of management innovation,

Table 2.3 Research perspectives on management innovation

	Condition perspective	Effectiveness perspective	Diffusion perspective	Process perspective
Core question	What conditions give rise to management innovation?	How can improve the performance of management innovation?	How can new management ideas be delivered?	How can management innovation occur?
Major research	Guillén (1994), Kossek (1987), McCabe (2002), Howell and Higgins (1990), Osterman (1994), Chi et al. (2007)	Stata (1989), Garvin (1993), Barker (1999), Li (2007), Teece (1986) Teece et al. (1997), Zahra and George (2002), Eisenhardt and Martin (2000)	Abrahamson (1996), Teece (1980), Carson et al. (2000), Rogers(1995)	Ahire and Ravichandran (2001), Cooper and Zmud (1990), Grover and Goslar (1993), Rogers (1995), Hamel (2006), Birkinshaw et al. (2007, 2008)

mechanism for delivering new management ideas and the process through which management innovation occurs respectively. Though great achievements have been made, the literature focusing on adoptive management innovation contains gaps. Most research on conditions and effectiveness perspective offer no preference to generating or adoptive type of management innovation (e.g. McCabe 2002; Howell and Higgins 1990). Moreover, proponents of diffusion perspective pay more attention to adoptive type by focusing on the diffusion of specific practices across firms or the boundary, but not its internal generative mechanism at the organization level (e.g. Abrahamson 1996; Burns and Wholey 1993; Teece 1980).

Research on process perspective stresses technological innovation (e.g. Rogers 1995; Pierce and Delbecq 1977) or generating type (e.g. Birkinshaw and Mol 2006; Birkinshaw et al. 2007, 2008), rather than adoptive type. Vaccaro et al. (2012) and Mol and Birkinshaw (2009) turn their attention directly towards adoptive management innovation by investigating the role of leadership behavior and demonstrating a trade-off between context and search respectively, little research goes into exploring how adoptive management innovation comes about. As a systematical project with high risks, adoption refers to advancing a new frame based on existing problems and resources of a certain organization (Mol and Birkinshaw 2009), and its success relies on the fit between existing practices and the new organizational environment. Therefore, it is a complicate process needing to be further explored, especially under the special context of China where firms are not developed enough to meet the demands for implementing various new management practices. Consequently, this research first seeks to address this gap by investigating the phases of adoptive management innovation and focusing on roles of core managers or internal agents.

Research literatures, particularly those on the rationality theory, stress the important role of individual managers in adopting and implementing new management practices. They argue that it is managers that address critical problems by creating or adopting new practices and offering support for implementation (Howell and Higgins 1990). However, little attention has been paid to how these individual managers make the decision of adoption or why some managers tend to adopt new practices. Therefore, based on the framework set up through a case study, this research further clarifies the process of adoption decision and examines how individual characteristics affect management innovation adoption level of a firm.

Though implementation has been identified as an indispensable phase without which firms are unable to realize the value of innovations (e.g. Birkinshaw et al. 2008; Lin and Su 2014), most research consider it as a natural action which does not need much attention. However, the intention of management innovation is to recreate routines of an organization (McCabe 2002). Or, management innovation potentially require fundamental changes in the routines of an organization, which reflects the difficulty and complication of innovation implementation (Argyris and Schän 1978; Birkinshaw et al. 2008). That is, the implementation of management innovation refers to a project of overturning existing organizational routines and recreating new ones, which deeply involves the most micro level of an organization. Therefore, this research attempts to explore the implementation mechanism of management innovation through an in-depth investigation into the micro-level of organizational routines, thereby addressing two questions: how firms implement management innovation through recreation of organizational routines? And how organizational routines evolve in an organization?

Research literatures on effectiveness perspective emphasize the relationship between dynamic capability and innovation, and contend that a firm's dynamic capability could significantly enhance its ability to innovate (O'Connor 2008), especially in the case of radical management innovation. That is, management innovation is not an isolated project, but relies on dynamic capability of a firm. Even though, literatures still contain gaps in discussing how dynamic capability internally enhances performance of management innovation. A close look at the body of literature also reveals that a majority of research are theoretical and conceptual. In consistence with these studies suggesting that dynamic capability is helpful in enacting changes in a systematic and fruitful way, this research attempts to explore how different components of dynamic capability affect each phase of adoptive management innovation, following the setting up of a process model for innovation.

Additionally, how management innovation affects organizations remains an unaddressed issue, which needs an urgent intention. This research argues that management innovation may produce effects on organizations through product innovation. It attempts to identify major management innovation conducted in the innovation process of complex product with a separate brand and explores how these kinds of management innovation support different phases of product innovation. To do so, a deep investigation into the whole innovation process of Lexus of Toyota Motor was conducted, aiming to find out all management innovation activities and analyze their effects. Based on this, it further explores how management innovation supports each phase of product innovation to produce the alignment effect.

In conclusion, as Table 2.4 shows, this research aims to address five major issues about adoptive management innovation: What is the process through which adoptive management innovation occurs? How do core managers of a firm make

Table 2.4 Research issues and research methods

Research issues	Research methods
What is the process through which adoptive management innovation occurs?	Case study
How do core managers of a firm make the decision of adopting new management practices to improving its efficiency?	Case study and CB-SEM
How could management innovation be implemented deeply at the most micro level of organizations, namely organizational routines?	Case study
How could adoptive management innovation be internally driven by dynamic capability?	Case study and PLS-SEM
How could intangible (adoptive) management innovation with ambiguous outcomes affect organizations through tangible technology innovation?	Case study

the decision of adopting new management practices to improving its efficiency? How could management innovation be implemented deeply at the most micro level of organizations, namely organizational routines? How could adoptive management innovation be internally driven by dynamic capability? And how could intangible (adoptive) management innovation with ambiguous outcomes affect organizations through tangible technology innovation? This research focuses on these five issues not only because of existing gaps in research, but also because they are concerning success of adoptive management innovation when investigating into abundant innovation cases especially from Chinese firms.

When addressing these issues, case studies, especially longitudinal interpretive and exploratory case studies, were adopted as the main research method to exert their function of theories construction (Eisenhardt 1989, 1991; Yin 1994; Eisenhardt and Graebner 2007), which was consistent with the process orientation of the study. The main purpose to adopt case study method is to combine literature to extract basic research framework and conduct initial inspection. According to Yin (1994), case studies are effective for theory development as they are rich empirical descriptions of particular instances of a phenomenon, or emphasize the rich, real-world context in which the phenomena occurs and forms the basis for inductive theory development (Eisenhardt and Graebner 2007; Yin 1994). Its essence is to derive research conclusions or new research propositions through describing what the phenomenon is in detail, analyzing its reasons, and finding or exploring the general rules and particularities, that is, seeing the hidden underlying factors through the appearance of things. Then, the central notion of adopting a case study is to use cases as the basis for developing theory inductively (Eisenhardt and Graebner 2007). As Eisenhardt and Graebner (2007, p. 25) argue, "a major reason for the popularity and relevance of theory building from case studies is that it is one of the best (if not the best) of the bridges from rich qualitative evidence to mainstream deductive research."

Since the theory-building process is deeply embedded in rich empirical data and occurs via recursive cycling among the case data, emerging theory, and later, extant literature (Eisenhardt 1989; Eisenhardt and Graebner 2007), and building theory from cases is likely to produce new theories that are accurate, interesting, and testable (Eisenhardt and Graebner 2007). Or, the theory is emergent in the sense that it is situated in and developed by recognizing patterns of relationships among constructs within and across cases and their underlying logical arguments. Although sometimes seen as "subjective", a well-done theory building from cases is surprisingly "objective", because its close adherence to the data keeps researchers "honest". More specifically, the conclusion of case study comes directly from enterprise practices or empirical evidences, so it can reflect reality more objectively. The data provide the discipline that mathematics does in formal analytic modeling. Research on management innovation stresses that, after summing up the enterprise management innovation practices, the objective facts in enterprise practices are explored, the features or conclusions are extracted and applied to more enterprise practices, reaching the purpose of improving management innovation theory and the actual performance level of the enterprise. Hence, case studies have been adopted by more and more researchers to develop various theories in organization research, especially where basic theoretical exploration lacks, for example, research on middle manager sense-giving (Balogun and Johnson 2004), organizational ambidexterity (Andriopoulos and Lewis 2009), and organizational identity change (Clark et al. 2010).

Since not only gaps do exist in the literature of adoptive management innovation but also little research goes into exploring issues as how adoptive management innovation comes about, how innovation should be implemented, or how management innovation may affect performance of an organization, exploratory case studies might be appropriate for developing theories of adoptive management innovation. Additionally, as Yin (1994) suggests, when addressing questions of "how" or "why", the case study would be more suitable. In this research, all concerns with complicate "how" or "why" questions. In addition, the case study often adopts some qualitative methods such as induction and deduction, and the theory obtained through this process has already passed the logical deduction so that the conclusion is more reliable.

As for the number of cases, Eisenhardt (1989), Eisenhardt and Graebner (2007) advocate the multiple-case design, by considering the replication of logic that each case serves as a distinct experiment. That is, multiple cases are discrete experiments that serve as replications, contrasts, and extensions to the emerging theory (Yin 1994). However, a single case is also useful where it represents a critical case, where it is an extreme or unique case, or where it is a revelatory case (Yin 1994). For example, when examining "sensemaking" during an imposed shift from hierarchical to decentralized organizations, Balogun and Johnson (2004) present a longitudinal, qualitative study on the single case of a privatized utility in the United Kingdom that was implementing planned strategic changes. Similarly, when exploring changes and managerial sensemaking, Lüscher and Lewis (2008) focus on the single case of Danish Lego Company and conduct action researches to construct a process of working through paradox.

Accordingly, case studies were adopted to construct a theoretical system of adoptive management innovation. More specifically, a single case study on the Organizational Efficiency Management of Jiangxi Mobile in China was adopted to develop the process framework of innovation in Chap. 3; a single case study on the case of Day-Definite (DD) innovation which successfully brought Arima World Group Company Limited (HOAU) into a new value-added arena, in terms of timing, security, and high service quality, was adopted to explore the complex implementation mechanism of management innovation in Chap. 5; a single case study on the market-chain-based Business Process Reengineering of Haier Group in China was adopted to redevelop a framework of adoptive management innovation before the exploration of how dynamic capabilities affect the process of innovation in Chap. 6; and also, a longitudinal and single interpretive and exploratory case study on Lexus of Toyota Motor was adopted to set up a general framework that explicates support and promotion mechanism of management innovation to the process of complex product innovation in Chap. 7. A single case study was adopted in these chapters because the chosen case was a critical one that meet all the necessary conditions for exploring certain issues on adoptive management innovation. Moreover, it allowed us to investigate the phenomena in depth to provide rich descriptions and understandings (Walsham 1995) of certain adoptive management innovation practices. Additionally, a multiple-case study was adopted to set up the framework of this research on adoptive management innovation and find out major issues, and to reset up the decision-making framework of adoptive management innovation in Chap. 4 before examining effects of the three dimensions on innovation decision making.

Besides case studies, this research also adopted literature research method and Structural Equation Modeling (SEM). Literature research method was used to select specific research issues and to offer theoretical background for the whole research and also for each separate issue. SEM, a diverse set of mathematical models, computer algorithms, and statistical methods that fit networks of constructs to data, or a statistical technique for testing theoretical models or proposed hypotheses, was adopted to confirm the set up frameworks of innovation decision making and effects of dynamic capabilities on innovation. Covariance-Based SEM (CB-SEM) and the software of AMOS 17.0 were adopted to examine the roles of the three dimensions of core managers in innovation decision making; while Partial Least Squares SEM (PLS-SEM) and Smart-PLS 2.0 by Ringle et al. (2005) were used to confirm the effects of dynamic capabilities on innovation. CB-SEM estimates model parameters so that the discrepancy between the estimated and sample covariance matrices is minimized. In contrast, PLS-SEM maximizes the explained variance of the endogenous latent variables by estimating partial model relationships in an iterative sequence of ordinary least squares (OLS) regressions. An important characteristic of PLS-SEM is that it estimates latent variable scores as exact linear combinations of their associated manifest variables and treats them as perfect substitutes for the manifest variables. The scores thus capture the variance that is useful for explaining the endogenous latent variable(s). Estimating models via a series of OLS regressions implies that PLS-SEM relaxes the assumption of multivariate normality needed for maximum likelihood–based on SEM estimations (Hair et al. 2012). That is why PLS-SEM rather than CB-SEM was adopted when the sample number was not large enough to get accepted goodness-of-fit statistics. The research issues and research methods are summarized in Table 2.4.

References

Abrahamson, E.: Management fashion. Acad. Manag. Rev. **21**(1), 254–285 (1996)

Ahire, S.L., Ravichandran, T.: An innovation diffusion model of TQM implementation. IEEE Trans. Eng. Manage. **48**(4), 445–464 (2001)

Andriopoulos, C., Lewis, M.W.: Exploitation-exploration tensions and organizational ambidexterity: managing paradoxes of innovation. Organ. Sci. **20**(4), 696–717 (2009)

Angle, H.L., Ven, A.H.V.D.: Suggestions for managing the innovation journey. In: Van de Ven A. H., Angle, H.L., Poole M.S. (eds.) Research on the Management of Innovation: The Minnesota Studies, pp. 663–697 (2000)

Ansari, S.M., Fiss, P.C., Zajac, E.J.: Made to fit: how practices vary as they diffuse. Acad. Manag. Rev. **35**(1), 67–92 (2010)
Argyris, C., Schön, D.A.: Organizational Learning: A Theory of Action Perspective. Birkhäuser, Basel (1978)
Balogun, J., Johnson, G.: Organizational restructuring and middle manager sensemaking. Acad. Manag. J. **47**(4), 523–549 (2004)
Barker, J.: The discipline of teamwork: participation and concertive control. Sage, Thousand Oaks, CA (1999)
Barley, S.R., Kunda, G.: Design and devotion: surges of rational and normative ideologies of control in managerial discourse. Adm. Sci. Q. **37**(3), 363–399 (1992)
Birkinshaw, J., Crainer, S., Mol, M.: Special report on management innovation. Bus. Strategy Rev **18**(1), 45–73 (2007)
Birkinshaw, J., Hamel, G., Mol, M.J.: Management innovation. Acad. Manag. Rev. **33**(4) (2008)
Birkinshaw, J., Mol, M.J.: How management innovation happens. Mit Sloan Manage. Rev. **47**(4), 81–88 (2006)
Burgelman, R.A.: A model of the interaction of strategic behavior, corporate context, and the concept of strategy. Acad. Manag. Rev. **8**(1), 61–70 (1983)
Burns, L.R., Wholey, D.R.: Adoption and abandonment of matrix management programs: effects of organizational characteristics and interorganizational networks. Acad. Manag. J. Acad. Manag. **36**(1), 106–138 (1993)
Caroli, E., Reenen, J.V.: Skill-biased organizational change? Evidence From a panel of British and French establishments. Cepremap Working Pap. **116**(4), 1449–1492 (2001)
Carson, P.P., Lanier, P.A., Carson, K.D.: Clearing a path through the management fashion jungle. Acad. Manag. J. **43**(6), 1143–1158 (2000)
Chi, W., Freeman, R.B., Kleiner, M.M.: Adoption and termination of employee involvement programs. NBER Working Paper, W12878 (2007)
Clark, S.M., Gioia, D.A., Ketchen, D.J., Thomas, J.B.: Transitional identity as a facilitator of organizational identity change during a merger. Adm. Sci. Q. **55**(3), 397–438 (2010)
Cole, R.E.: The macropolitics of organizational change: a comparative analysis of the spread of small-group activities. Adm. Sci. Q. **30**(4), 560–585 (1985)
Cooper, R.B., Zmud, R.W.: Information technology implementation research: a technological diffusion approach. Manage. Sci. **36**(2), 123–139 (1990)
Damanpour, F., Schneider, M.: Phases of the adoption of innovation in organizations: effects of environment, organization and top managers. Br. J. Manag. **17**(3), 215–236 (2006)
Edmondson, A.: Methodological fit in management research. Plast. Reconstr. Surg. **108**(3) (2003)
Edmondson, A., Bohmer, R., Pisano, G.: Speeding up team learning. Harvard Bus. Rev. **79**(9), 125–132 (2001)
Eisenhardt, K.M.: Building theories from case study research. Acad. Manag. Rev. **14**(4), 532–550 (1989)
Eisenhardt, K.M.: Better stories and better constructs: the case for rigor and comparative logic. Acad. Manag. Rev. **16**(3), 620–627 (1991)
Eisenhardt, K.M., Graebner, M.E.: Theory building from cases: opportunities and challenges. Acad. Manag. J. **50**(1), 25–32 (2007)
Eisenhardt, K.M., Martin, J.A.: Dynamic capabilities: what are they? Strateg. Manag. J. **21** (2000)
Garvin, D.A.: Building a learning organization. Harvard Bus. Rev. **71**(4), 78–91 (1993)
Grover, V., Goslar, M.D.: The initiation, adoption, and implementation of telecommunications technologies in U.S. organizations. J. Manag. Inf. Syst. **10**(1), 141–163 (1993)
Guillén, M.F.: Models of management: work, authority, and organization in a comparative perspective. University of Chicago Press, Chicago (1994)
Hage, J.: Theories of Organization: Form, Process, and Transformation. Wiley, New York (1980)
Hair, J.F., Sarstedt, M., Ringle, C.M., Mena, J.A.: An assessment of the use of partial least squares structural equation modeling in marketing research. J. Acad. Mark. Sci. **40**(3), 414–433 (2012)
Hamel, G.: The why, what, and how of management innovation. Harvard Bus. Rev. **84**(2), 72–84 (2006)

Howell, J.M., Higgins, C.A.: Champions of technological innovation. Adm. Sci. Q. **35**(2), 317–341 (1990)
Kimberly, J.R., Evanisko, M.J.: Organizational innovation: the influence of individual, organizational, and contextual factors on hospital adoption of technological and administrative innovations. Acad. Manag. J. Acad. Manag. **24**(4), 689–713 (1981)
Klein, K.J., Sorra, J.S.: The challenge of innovation implementation. Acad. Manag. Rev. **21**(4), 1005–1081 (1996)
Knights, D., McCabe, D.: 'Ain't Misbehavin'? Opportunities for resistance under new forms of 'quality management'. Sociology **34**(3), 421–436 (2000)
Kohlbacher, M.: The impact of dynamic capabilities through continuous improvement on innovation: the role of business process orientation. Knowl. Process Manag. **20**(2), 71–76 (2013)
Kossek, E.E.: Human resources management innovation. Hum. Resour. Manag. **26**(1), 71–92 (1987)
Kramer, H.: The philosophical foundations of management rediscovered. Manag. Int. Rev. **15**(2–3), 47–55 (1975)
Lazonick, W., Prencipe, A.: Dynamic capabilities and sustained innovation: strategic control and financial commitment at Rolls-Royce plc. Ind. Corp. Change **14**(3), 501–542 (2005)
Leibenstein, H.: Bandwagon, Snob, and Veblen effects in the theory of consumers' demand. Quart. J. Econ. **64**(2), 183–207 (1950)
Li, Y.: Management Innovation in Organizational Learning. Economic Management Press (2007)
Liao, J., Kickul, J.R., Ma, H.: Organizational dynamic capability and innovation: an empirical examination of internet firms. J. Small Bus. Manag. **47**(3), 263–286 (2009)
Lin, H.F., Su, J.Q.: Management innovation mechanism: improving the effectiveness of organizational ambidexterity perspective. Sci. Res. Manag. **33**(2), 1–10 (2012)
Lin, H.F., Su, J.Q.: A case study on adoptive management innovation in China. J. Organ. Change Manag. **27**(1), 83–114 (2014)
Lüscher, S.L., Lewis, M.W.: Organizational change and managerial sensemaking: working through paradox. Acad. Manag. J. **51**(2), 221–240 (2008)
Lynch, L.M.: The adoption and diffusion of organizational innovation: evidence for the U.S. economy. In: IZA Discussion Paper, 5, No. 2819 (2007)
March, J.G.: Exploration and exploitation in organizational learning. Organ. Sci. **2**(1), 71–87 (1991)
Mccabe, D.: 'Waiting for dead men's shoes': towards a cultural understanding of management innovation. Hum. Relat. **55**(5), 505–536 (2002)
Mccabe, D., Knights, D.: A road less travelled: beyond managerialist, critical and processual approaches to total quality management. J. Organ. Change Manag. **15**(3), 235–254 (2002)
Mol, M.J., Birkinshaw, J.: The sources of management innovation: when firms introduce new management practices. J. Bus. Res. **62**(12), 1269–1280 (2009)
Naveh, E., Meilich, O., Marcus, A.: The effects of administrative innovation implementation on performance: an organizational learning approach. Strategic Organ. **4**(3), 275–302 (2006)
O'Connor, G.C.: Major innovation as a dynamic capability: a systems approach. J. Prod. Innov. Manag. **25**(4), 313–330 (2008)
Osterman, P.: How common is workplace transformation and who adopts it? Ind. Labor Relat. Rev. **47**(2), 173–188 (1994)
Pierce, J.L., Delbecq, A.L.: Organization structure, individual attitudes and innovation. Acad. Manag. Rev. **2**(1), 27–37 (1977)
Piva, M., Vivarelli, M.: The determinants of the skill bias in Italy: R&D, organisation or globalisation? Econ. Innov. New Technol. **13**(4), 329–347 (2004)
Ringle, C.M., Wende, S., Will, A.: SmartPLS 2.0. University of Hamburg, Hamburg, Germany (2005)
Rogers, E.M.: Diffusion of Innovations, 4th edn. The Free Press, New York (1995)
Stata, R.: Organizational learning-the key to management innovation. Mit Sloan Manag. Rev. **63** (1), 63–73 (1989)

Staw, B.M., Epstein, L.D.: What bandwagons bring: effects of popular management techniques on corporate performance, reputation, and CEO pay. Adm. Sci. Q. **45**(3), 523–556 (2000)

Stjernberg, T., Philips, A.: Organizational innovations in a long-term perspective: legitimacy and souls-of-fire as critical factors of change and viability. Hum. Relat. **46**(10), 1193–1219 (1993)

Suddaby, R., Greenwood, R.: Rhetorical strategies of legitimacy. Adm. Sci. Q. **50**(1), 35–67 (2005)

Teece, D.J.: The diffusion of an administrative innovation. Manage. Sci. **26**(5), 464–470 (1980)

Teece, D.J.: Profiting from technological innovation: implications for integration, collaboration, licensing and public policy. Res. Policy **15**(6), 285–305 (1986)

Teece, D.J.: Explicating dynamic capabilities: the nature and microfoundations of (sustainable) enterprise performance. Strateg. Manag. J. **28**(13), 1319–1350 (2007)

Teece, D.J., Pisano, G., Shuen, A.: Dynamic capabilities and strategic management. Strateg. Manag. J. **18**(7), 509–533 (1997)

Vaccaro, I.G., Jansen, J.J.P., Bosch, F.A.J.V.D., Volberda, H.W.: Management innovation and leadership: the moderating role of organizational size. J. Manag. Stud. **49**(1), 28–51 (2012)

Waarts, E., Everdingen, Y.M.V., Hillegersberg, J.V.: The dynamics of factors affecting the adoption of innovations. J. Prod. Innov. Manag. **19**(6), 412–423 (2002)

Walsham, G.: Interpretive case studies in is research: nature and method. Eur. J. Inf. Syst. **4**, 74–81 (1995)

West, M.A., Anderson, N.R.: Innovation in top management teams. J. Appl. Psychol. **81**(6), 680–693 (1996)

Wu, Q., He, Q., Duan, Y., O'Regan, N.: Implementing dynamic capabilities for corporate strategic change toward sustainability. Strateg. Change Spec. Issue Entrepreneurship Innov. **21**(5–6), 231–247 (2012)

Yin, R.K.: Case Study Research: Design and Methods, 2nd edn. Sage, Newbury Park, CA (1994)

Zahra, S.A., George, G.: Absorptive capacity: a review, reconceptualization, and extension. Acad. Manag. Rev. **27**(2), 185–203 (2002)

Zaltman, G., Duncun, R., Holbek, J.: Innovations and Organizations. Wiley, New York (1973)

Zbaracki, M.J.: The rhetoric and reality of total quality management. Adm. Sci. Q. **43**(3), 602–636 (1998)

Zhang, G.: The learning mechanism and knowledge management in the process of firm's organizational innovation. Sci. Res. Manag. **20**(3), 40–44 (1999)

Chapter 3
What Is the Process Through Which Adoptive Management Innovation Occurs?

3.1 Introduction

As a systematical project with high risks, adoption refers to advancing a new frame based on existing problems and resources of a certain organization (Mol and Birkinshaw 2009), and the success of adoptive management innovation relies on the fit between existing practices and the new organizational environment. Therefore, it is a complicate process needing to be further explored, especially under the special context of China where firms are not developed enough to meet the demands for implementing various new management practices. Additionally, as a big issue concerning overall benefits and future development of organizations, the process of adoptive management innovation depends on core managers who are responsible for making essential decisions and controlling overall operations of the company (Su and Lin 2010). Consequently, this chapter seeks to address this gap by investigating the phases of adoptive management innovation and focusing on roles of core managers or internal agents.

More specifically, this chapter focuses on the generative mechanism of adoptive management innovation and explores different phases by considering the question of how adoptive management innovation occurs. Since the nature of adoptive management innovation is different from that of generative innovation, it may go through a different process from that of new-to-the-state-of-the-art management innovation proposed by Hamel (2006) and Birkinshaw et al. (2008). Furthermore, adoption means advancing a new scheme by integrating existing practices into problems and resources of the new organization (Mol and Birkinshaw 2009) rather than imitating or repeating practices of other organizations. In a word, this chapter aims to address how management practices implemented successfully somewhere else have been introduced into firms and implemented effectively, or to investigate key activities of the generative mechanism through which an adoptive management innovation occurs.

H. Lin, *Adoptive Management Innovation*,
https://doi.org/10.1007/978-981-10-7712-8_3

This chapter is organized as follows. Section 3.2 presents the method adopted in the research including why the case study method adopted and how data is collected and analyzed. Section 3.3 shows the findings obtained from the case analysis on Organization Efficiency Management of Jiangxi Mobile in China. Section 3.4 develops a process framework of adoptive management innovation, and also discusses implications. Finally, Sect. 3.5 offers a brief conclusion.

3.2 Method

3.2.1 The Case Study

Exploratory case study refers to one of research methods of the problems, which summarizes from organization practices thus extracts theory models or transforms practices into theories, when the research theory is vacant or existing research is not rich enough to make a full explanation of the research question (Wu and Yu 2004). A single longitudinal interpretive and exploratory case study approach is adopted in this chapter so as to explore how adoptive management innovation occurs for the following reasons: firstly, the research on management innovation is still in the exploratory stage, especially for the related research on the problem of how management innovation occurs is more deficient, which makes the theoretical basis of this research topic vacant; secondly, how adoptive management innovation occurs is actually a complex and abstract process and needs to be further excavated, while, the case study is suitable for addressing the gaps in research on innovation process and the case chosen here is a critical one that meets all the necessary conditions for developing a process framework of innovation; thirdly, the case study can meet the demands of researchers to reveal the essence of management innovation and relevant mechanisms through exploring management innovation practices of an organization, meanwhile, it's able to better understand and find out the problems in practices by deeply contacting with the target enterprise and related individuals. Moreover, it allows us to investigate the phenomena in depth to provide rich descriptions and understandings (Walsham 1995) of certain adoptive management innovation practices.

3.2.2 Data Collection

The research setting was Jiangxi Mobile Communication Company Limited (Jiangxi Mobile), one of the best subsidiaries of China Mobile Group (the leading state-owned mobile services provider in Mainland China which boasts the world's largest mobile network and the world's largest mobile customer base, and was once again selected as one of the "FT Global 500" by Financial Times and "The World's

2000 Biggest Public Companies" by Forbes magazine, and was again recognized on the Dow Jones Sustainability Indexes in 2011). As illustrated in Table 3.1, Jiangxi Mobile was founded in 1999 with 11 city and 85 county branches; and it engaged in offering voice services (for example, local calls, domestic long distance calls, international long distance calls, intra-provincial roaming, inter-provincial roaming, international roaming and voice value-added services), data services (for example, SMS and MMS, Wireless Data Traffic, Applications and Information Services) and the customer service (i.e., nationwide free customer service hotline "10086") like most of other mobile services providers. In recent years, Jiangxi mobile has gone through a rapid-developing period and consolidated its dominant position in the regional communications market, which could be attributed to the efforts made to introduce and implement various management practices or methods in response to environmental changes in its competitive market, including Strategic Management, Budget Management, Performance Management and Organization Efficiency Management. Here the research focused on one of these innovations, namely, Organization Efficiency Management, and its realizing process (2004–2007), to explore how Jiangxi Mobile successfully adopted and implemented this new practice deeply and systematically, in order to develop a process framework.

The case of Organization Efficiency Management is adopted for two reasons: one is that Organization Efficiency Management represents a typical adoptive management innovation which has been implemented successfully in various firms around the world especially in western countries. The efficiency management thought has been widely used in western communications industry, its core technology is the eTOM (enhanced Telecom Operations Map), which is the business process framework of the information and communication service industry. The eTOM describes requirements from the perspective of business view, analyzes and designs the business process, then forms solutions after systematically analyzing and designing, ultimately puts in actual operations and meets customers' demands after passing the conformance testing of solutions. With the development of eTOM business model, the model included in the knowledgebase can become a demand for interactive resource. The processes and flows decomposed in the model can be directly connected to the system and the implementation of components, in order to meet the requirements of the business process. Based on the experience from foreign companies in implementing organization efficiency management and its key technique eTOM, Jiangxi Mobile adopted this new management method in 2004 and successfully implemented it by integrating the new concept into its specific context. The other is that, until 2009 (five years after the beginning of adopting organization efficiency management innovation), Jiangxi Mobile had made a series of great achievements: (1) The response time of serving customers' demands was shortened, comprehensive operation efficiency and the construction of excellent operating system were promoted, for example, the customer satisfaction rate had been raised from 72.2% in 2006 to 80.5% in 2009, and the customer complaint handling time had been shortened from 19.5 h in 2006 to 8 h in 2009. (2) More importantly, employees had been encouraged to work cohesively to support the company's new strategy through this innovation, for example, the ability of

Table 3.1 Overview of the case company

Company	Belonging	Owner-ship	Service or products	Year founded	Branches	Management innovation	The CASE innovation
China Mobile Group Jiangxi Company Limited (Jiangxi Mobile)	A subsidiary of China Mobile Group	State-owned	Voice services, data services and the customer service	1999.8	11 City and 85 county branches	For example, strategic management, budget management, performance management and organization efficiency management	Organization efficiency management (2004–2007)

company organization to response market was 85%, reaching the level of excellence, and the indicators, including "direction of the enterprise", "leadership" and "incentive reward mechanism", were raised from the normal level in 2006 to the excellent level. (3) The implementation of new cross strategy was effectively supported to build the enterprise sustainable development ability. (4) A set of shareable and reproducible system methods and simple and practical tools were formed, which provided references for China Mobile Group and other organizations to carry out the Efficiency Management. Consequently, a set of systematic methods for improving the operation efficiency of organizations were finally developed, which had brought abundant insights into further development of the whole group of China Mobile and other organizations. In a word, the implementation of Organization Efficiency Management in Jiangxi Mobile was of a successful case of adoptive management innovation in China with a strong research value.

Multiple approaches were used during data collection to meet criteria for trustworthiness (Lincoln and Guba 1985; Yin 1994), including semi-structured interviews, archival data, and observation.

Semi-structured interviews. Three sets of interviews were conducted in this research, as shown in Table 3.2. The first set of interviews with two members of the top-level manager group (the general manager and a vice general manager) in Jiangxi Mobile were conducted to get some basic information about this company on December 14th and 15th, 2009 (for example, the history of the company, its missions, evolution of management model, characteristics of management, culture for innovations, main management innovation practices in the past ten years, background and antecedents for Organization Efficiency Management, context for this innovation, and its general process and outcomes). They were interviewed two times in two or three hours respectively. The second set of 10 semi-structured interviews occurred over a two-month period from February to March 2010, with 9 top-level and middle-level managers (recommended by the two top managers) including another vice general manager who was directly in charge of the

Table 3.2 Quantitative details of interview data

Interviews	Informant	Time	Purpose
The first set of interviews: 2 interviews	2 Members of the top-level manager group: the general manager and a vice general manager	December 14 and 15, 2009; two and three hours respectively	To get some basic information about this company
The second set of interviews: 10 semi-structured interviews	9 Top-level managers and department managers	From February to March 2010; ranged 60–90 min	To get their impressions and cognitions on this management innovation process
The third set of interviews: 24 informal interviews	First-line managers and employees who participated in the implementation process	April 2010; within 60 min	To generate additional information and provide a better understanding of the innovation

innovation practice (two interviews) and department managers of Strategic Development, Human Resources Management, Marketing, Planning Development, Financial, Service and Products, Part Affairs and Network (one interview respectively) involving in the innovation. This set of interviews was about their impressions and cognitions on this management innovation process, which offered both general information and some details about the whole adoption and implementation process of Organization Efficiency Management (for example, main themes and issues in the process, key managers and participants in each key phase, roles of core managers in each phase, activities and responsibilities of each department, and contributions and achievements). These interviews ranged 60–90 min in length. Additionally, in order to fully understand the information, a third set of informal interviews within 60 min with some first-line managers and employees (a total of 24 interviews) who participated in the implementation process was conducted to generate additional information and to provide a better understanding of the innovation in April, 2010. Interviews, especially these with top managers, began with questions covering more general topics. In view of the inductive aims, the informants were encouraged to go deeply into the details and internal key links of the innovation. Before each set of interviews, an interview protocol was designed with major themes in mind based on the basis of a preliminary understanding of innovation events; and during the interviews, questions were not asked in any specific order but were governed instead by the actual situation of the informants (Gummesson 2000), and some additional problems might be proposed to deeply excavate more valuable information and find out some hidden details.

Archival data. In addition to the interview data, the author also collected archival data in forms of the innovation proposal, relevant meeting notes, newsletters, memos, annual reports and innovation performance reports. The strategic development department who was in charge of the whole innovation events gave the access to all minutes of meetings and memos relevant to the innovation, such as the records of the innovation process and the evaluations of the effects of Efficiency Management in recent years. The marketing department offered the marketing responding results of the innovation, for example, the increased customer satisfaction rate and the shortened customer complaint handling time. Moreover, the author gathered various articles, media reports, stories and Web materials related to Jiangxi Mobile and its Organizational Efficiency Innovation.

Observation. Since this research was conducted after the innovation had been finished, the author could not observe its specific process of adoption and implementation. Fortunately, the author got the opportunity to attend the meeting for post appraising the effectiveness of this innovation in the end of 2009 (two years after it was finished) and took notes about comments on its process and outcomes. Furthermore, during each visit, notes about informal observations while waiting for interviews and working around the company were made. The service halls, where customers were offered with various services directly, offered the opportunity to observe interactions between employees and customers which indicated the improved quality of operation and service after the innovation. All of these observation notes relevant to the implementation effects of the Efficiency Management were considered as the supplementary data to enrich the case data.

3.2.3 Data Analysis

The analysis, guided by Eisenhardt's notion that "it is the intimate connection with empirical reality that permits the development of a testable, relevant, valid theory" (1989, p. 532), proceeded in five stages. Systematic, iterative comparisons of data, emerging categories, and existing literature aided development of cohesive constructs and an integrative, theoretical framework of adoptive management innovation. Firstly, after being collected via approaches of interviews, archival data and observation, the data was collated and sorted before analysis, for example, interviews were taped and transcribed, field notes were collated and observations were written up, with rich raw data transformed into the written form. Specifically, Semi-structured interviews were the primary source of the data, which offered the most information about how the process of Efficiency Management in Jiangxi Mobile was realized; and archival materials and observations expanded the understandings of the case by offering insights that might refute or reinforce the interview findings (Andriopoulos and Lewis 2009). Secondly, via blending all data together and taking it into a sequential order, the author wrote a story of Organization Efficiency Management in Jiangxi Mobile. The intention was to capture the ebb and flow of change interventions and activities as well as the interpretations of managers and employees on these events. In the third stage of analysis, key activities in the whole sequential process and tasks of key managers in each phase were identified by adhering to guidelines specified for methods of naturalistic inquiry and constant comparison techniques (Lincoln and Guba 1985; Strauss and Corbin 1990). More specifically, the author and another team member coded the case story separately on the basis of in vivo phrases used by the informants, and then compared the coding results with each other to reach a consensus on the main activities and links in the whole process of innovation. The codes discerned similarly were collated into first-order activities or categories. Fourthly, after developing the first-order activities, the author started discerning linkages among the activities that could lead to the development of second-order themes by formulating researcher-induced concepts at a more abstract level. Finally, the author assembled the second-order themes into aggregate phases or sub-processes to develop a process framework.

3.3 Findings

Figure 3.1 shows the data structure of the findings. It depicts the two main phases or sub-processes of adoption decision and implementation in the process of adoptive management innovation and their first-order activities or categories (which are directly concluded based on the acquired data coding, for example, analyzing Symptoms of "large company disease", realizing the application of 3G and reorganization of the whole telecoms industry, searching for a solution, focusing on Efficiency Management, initiatively evaluating Efficiency Management, investigating into the causes for low efficiency) and second-order themes (which are the further integration and summary of the former, for example, problem identification, innovation perception, attitude formation, problem diagnoses, innovation revision, proposal evaluation and selection, preparation for implementation, overall implementation, and solidification). This section will discuss the findings by exploring deeply into each phase of the whole Efficiency Management process that could be assembled into two general and interlinked sub-processes of adoption decision and implementation. The ensuing sections will expatiate on the whole process of innovation.

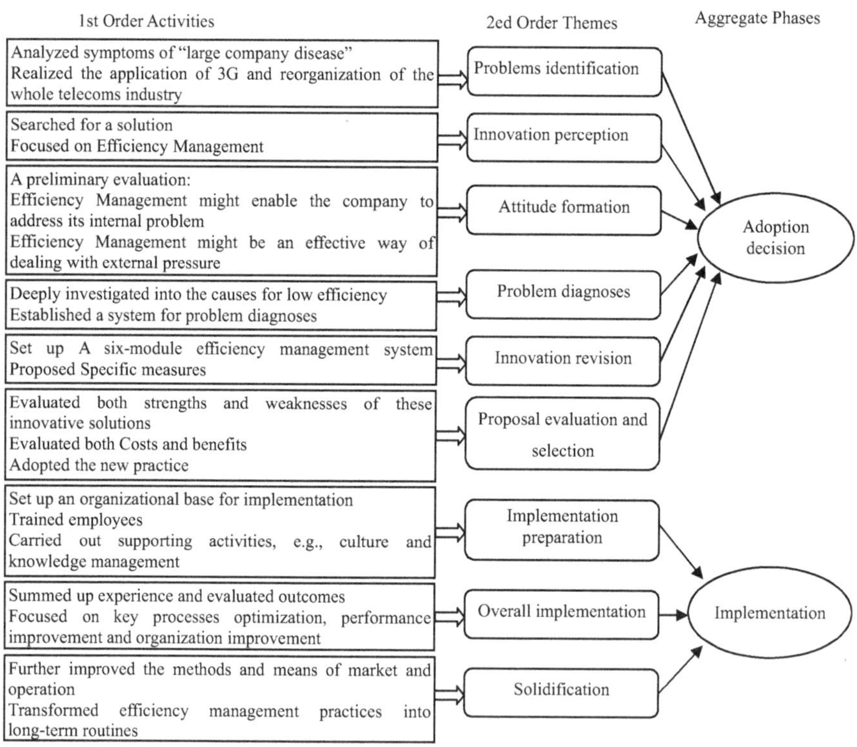

Fig. 3.1 Data structure

3.3.1 Adoption Decision

Adoption decision consists of themes of problem identification, innovation perception, attitude formation, problem diagnoses, innovation revisions, proposal evaluation and selection. In another word, it focuses on initiation and realization of the fitness between existing management practices and the new organizational context. Furthermore, it finds that top managers (especially the general manager) play a most important role in this adoption decision process.

Problem identification. Efficiency Management innovation began with initial identification of internal existing problems in Jiangxi Mobile and external challenges that it faced. From the internal view of point, as a state-owned company with large scale but low flexibility, Jiangxi Mobile had shown more and more symptoms of "large company disease":

> Recent years, problems of overstaffing, overlapping, low delivering speed of information, pool coordination, excessive growth of bureaucracy and low enthusiasm in innovation have arisen. This undoubtedly would keep our company from developing. (The general manager, interview)

Besides, the development of enterprise strategy needed to be supported. In order to actively adapt to the new requirements, new competition, new environment, headquarters of China Mobile established new strategic goal of "being the world first-class enterprise", which required themselves to have the first-class management level and the operational efficiency around the world.

From the external view of point, mobile communication technology gradually developed into the 3G technology based on the first generation of simulation technology mobile-phones and the second generation of digital mobile-phones. The application of 3G technology was a great technological change, during which the improvement of technology and service demands required the corresponding improvement of service capacity, especially the improvement of service process efficiency. At the same time, in the whole telecommunications industry, in addition to the saturation of fixed-phone demands, mobile-phone, SMS, Internet and other related data and digital demand presented a steady growth trend, and highlighted some characteristics such as "more mobile than fixed", "more data than voice" and personalized, high-efficient and diversified services, which puts forward a new challenge to telecom operators. In order to deal with the competitive pressure from the coming 3G (3rd Generation) technology which allowed mobile phone customers to enjoy high-speed Internet-based services in telecom industry, Jiangxi Mobile, as a service provider, needed an improvement in its operation and management to offer services more efficiently with new technologies. Meanwhile, the government of China planned to take advantage of issuing 3G licenses to optimize the structure of competition among its telecom operators. According to the plan, three state-run businesses, China Telecom, China Unicom and China Mobile, would compete for the telecommunications market in China through a revolution and restructuring directed by Ministry of Information Industry (MII), Nationals Development and Reform Commissions (NDRC) and Minister of Finance. Then, all of three telecom service providers in China would be granted with 3G licenses of fixed-line and fixed-mobile business. That is to say, the monopolized position of China Mobile (for example, in 2007, China Mobile accounted for 49% of the whole sector's earnings and occupied 69% of the whole domestic market) was to be broken after the restructuring. In short, with the application of 3G and the reorganization of the telecom industry, Jiangxi Mobile, as a main subsidiary of China Mobile, might have to face uncertainty and risks in integration of telecom, radio and TV, and internet networks, and integration of value chains as well as those of business and means, when facing the trend of telecoms market homogeneity and the trend of technology, business as well as service homogeneity. Under this background, managers in Jiangxi Mobile realized that:

> Management would undoubtedly become an essential source of competitive advantages for firms in telecom industry, and it was an urgent task to improve its operation efficiency. (The vice general manager, interview)

Innovation perception. Being confronted with these problems and challenges, managers of Jiangxi Mobile attempted to look the inside and outside their organization for a solution. Constrained both by the pressure to conform to the norms of rationality of the organization's institutional field and by the costs of evaluating multiple competing offers, managers would often decide to adopt a solution or innovation that appeared to be most progressive and legitimate (Abrahamson 1996). Therefore, managers of Jiangxi Mobile tried to search for new management practices via accessing to various knowledge sources, e.g., internal sources (managers and employees within an organization), market sources (customers, suppliers, competitors, and consultants), and professional sources (industry bodies, professional associations, and trade fairs) (Mol and Birkinshaw 2009). Finally, they got the concept of Organization Efficiency Management from its international competitors and a consulting institution (market sources), and viewed this new concept as *"a strategy-driven management practice that focused on philosophies of pursuing process supervision, rigid quantification, flexible reaction and interlinks of different modules, or a method enabling efficient combination of various elements, links, management measures, means and tools, and finally enabling the address of existing problems." (The vice general manager, interview).*

Attitude formation. As a management method of improving the operation efficiency of an organization, Organization Efficiency Management involved activities of analyzing existing organizational processes, evaluating their efficiency and proposing approaches for efficiency improving. The core of Organization Efficiency Management was process reengineering and management. The eTOM (enhanced Telecom Operations Map), which described the full scope of business process required by a service provider and defined key elements and how they interacted, was the most widely accepted and used techniques for process reengineering and management in the telecommunication industry. Integrating identified problems with the new Efficiency Management practice, managers realized that, for one hand, Efficiency Management would enable the address of internal operation problems in Jiangxi Mobile and facilitate its high-speed development. It was concerned with offering a coexisting mechanism of both the large-scale effect and small-scale flexibility, emphasizing coordination among departments, constant supervising operational efficiency of various processes, and establishing a standardized management system with both vertical and horizontal integration. For the other hand, Efficiency Management would be an effective way of dealing with external pressure by making in-time actions, focusing on the needs of customers, providing with high-quality products and offering high-quality services. Consequently, managers of Jiangxi Mobile formed a positive attitude toward the new management practice of Efficiency Management and initiated a change.

Problem diagnoses. The formed positive attitude drove managers to further the innovation. In order to clear specific problems of the organization to better introduce and implement the efficiency management, under the guidance of top managers, Jiangxi Mobile launched a series of activities in problem diagnoses. First, they deeply investigated the roots of low efficiency of the existing management system, where a new and specific scheme would be located. A system for thorough diagnosing was established with three paths of Interview and Investigation, Process System Carding and Efficiency Indexes Benchmarking: (1) Interview and Investigation reflected a process of making in-depth interviews with managers of different levels, first-line employees as well as external clients, partners, and other stakeholders, with the purpose of listening to their voice of mind and seeking for the sources of low efficiency, which provided a base of process carding and diagnosing. A designed questionnaire on performance perception was released through the net for an anonymous survey to objectively do some secret diagnoses by evaluating the company's competency in market reflection, leadership, development orientation, belongingness of employees and motivation. In order to further uncover individual factors that affected efficiency, a time management research was carried out to provide improved time management methods according to staff of different levels. (2) Process System Carding referred to a process of searching for key processes that needed to be further improved and optimized by evaluating both effectiveness and feasibility through the eTOM model. More specifically, it included four steps: the first step was to modify the primary process framework based on the old eTOM model, and rearrange the secondary process framework combining with the actual situation of Jiangxi Mobile; the second step was to map with the improved eTOM model, based on the collection of the company's existing processes, which formed the panoramic map of process framework of Jiangxi Mobile to promote the satisfaction of internal and external customers; the third step was to find out processes needed to be optimized in the process system through synthetically considering processes from the aspects of systems, ideas, setting, execution and supervision; the final step was to filtrated the key processes needed to emphatically be optimized by grading and sorting the processes needed to be optimized from two aspects of implementing effect and maneuverability. (3) Efficiency Indexes Benchmarking was an action of selecting most important indexes that might represent the efficiency of the organization from perspectives of market, operation, organization and employees, by integrating experience from top-level global telecom operators into the practices of Jiangxi Mobile, as illustrated in Table 3.3. For example, market effect and inter-department market reaction speed, which could be measured by customer satisfaction and average launch time of new products respectively, were two most important indexes selected for evaluating the market efficiency. On this basis, by evaluating each index combined with the present situation of organizational efficiency in Jiangxi Mobile, this research made a conclusion that though three indexes (reaction speed for group customers' demand, internal customers satisfaction and transferring speed of certain key process) had reached the excellent level, there still existed five indexes that needed to be improved, including customer satisfaction, average launch time of new products, the complaint handling rate, the average complaint handling time and working enthusiasm and initiation. That was to say, this research identified three levels of the overall evaluation results, among which the market and operating efficiency remained to be promoted and the efficiency of organization and individuals achieved excellence. Hence, the focus of efficiency management was to find out the roots of poor efficiency on the market and operating.

The results of systematic diagnoses showed specific problems existing in systems of process management, performance management, quality management, risk management, organization management and IT supporting in Jiangxi Mobile. Firstly, problems in the process management system included a lack of certain important processes, lagging or non-solidification of some processes, a lack of customer-oriented process, over length and redundancy in some process chains, slow reaction to challenges from competitors, slow reaction of functional departments to business departments, not-in-time resources allocation, allocating resources according to interpersonal relationship or negotiation capability instead of

Table 3.3 Efficiency management indexes benchmarking

Levels	Key indexes benchmarking types	Examples	Measures	
			Time	Outcome
Market efficiency	Market effect	Customer satisfaction		√
	Inter-department market reaction speed	Average launch time of new products	√	
Operation efficiency	Customer cognition	The complaint handling rate		√
	Inter-department support	The average complaint handling time	√	
		Reaction speed for group customers' demand	√	
Organization and employees efficiency	Organizational transferring efficiency	Internal customers satisfaction		√
		Transferring speed of certain key process	√	
	Working efficiency of employees	Working enthusiasm and initiation		√

customers' demand, low efficiency in new business and products development caused by difficult horizontal integration, using same processes regardless of priorities and sizes of tasks and redundant documents on processes. Secondly, the performance management system were subjected to excessive key performance indexes (KPI) coming from the provincial general headquarters, contradiction among some indexes, over emphasis of most indexes on the demand of departments but not the general goal of the company, lower adjusting speed of KPI over that of resources allocation (for example, no resource would be allocated to support new business goals in every season). Thirdly, for the quality management system, problems mainly included low speed of handling complaints, poor customer satisfaction in handling complaints, and a lack of network maintenance leading to poor quality in network construction. Fourthly, the risk management system showed problems of lower working efficiency perceived by employees in implementing the Sarbanes-Oxley act, a lack of risk early-warning mechanism, and over attention paid to technological risk but little to market risk. Fifthly, many problems emerged in the organization management system too, for example, excessive management resources allocated to provincial level but little to county level, multiple head management over the first line employees that made them tied of handling it, different organizational structures in different levels (leading to ambiguous assignments of tasks), overlapping or blank processes, unclear functions of departments, a shortage of human resources in some key functional positions (leading to low speed of execution and process bottlenecks), a gap between the skills of employees and the requirements for their positions (leading to a lack of effective employees), and low employee working satisfaction and enthusiasm, the levels of which even varied in different places. Finally, the IT supporting system also needed a change because it lacked some important supporting platforms (for example, a database for new business development, a platform of provider appraisement and management), the systematical analyses and integrations of information (leading to increasing analysis time and thus missing chances), an electronic support to cross-department and cross-level operation, and whole-process supervision and real-time controlling (e.g., the implementation and tracking of demands of group clients generated in the city). Moreover, problems of the IT supporting system were also concerned with high dependence on manpower in information analysis and a low-running and unstable BOSS (Business Operations Support System), leading to insufficient supports of business that implemented successfully in the past.

A further investigation and analysis implied that it was the defects of the company's operation system, organization structure, human resources and cultural system that jointly leaded to these problems in different management systems, as simply illustrated in Table 3.4. More specifically, the shortcomings of the operation system referred to: weaknesses in the fundamental management, blank points still existing in some processes and the IT supporting system and lacking of integrality, over reliance on the functional departments but not the whole organization when setting up the operation system, an attempt to link existing processes together but not covering the general interest, imbalance among quality, risk and efficiency

Table 3.4 Sources and descriptions of efficiency problems in Jiangxi Mobile

Sources of problems	Descriptions
Operation system elements	• Weaknesses in the fundamental management • Blank points in some processes and the IT supporting system • Over reliance on the functional departments • An attempt to just link existing processes together • Imbalance among quality, risk and efficiency controlling, management capability and foundation of the organization • Low scientificity, practicality and systematicness
Organizational elements	• The pure functional organizational structure • The diversified structure and function of city and count branches
Human and culture elements	• Over emphasis on "discharge" • Over emphasis on "levels" and "bureaucracy" • Inefficient motivation and punishment mechanism • An absence of the sense of crisis

controlling, management capability and fundamental of the organization, and low scientificity, practicality and systematicness of the KPI design and assessment system which were urgent to be further improved. These efficiency problems also stemmed from the organizational structure, for example, on the one hand, the pure functional organizational structure could not meet the requirement of customer-centered horizontal integration, while on the other hand, the diversified structures and functions of cities and count branches leaded to a shortage of vertical integration in some areas. Furthermore, some human resources and culture elements might lead to efficiency problems. Horizontally, overemphasis on the concept of "discharge" leaded to dissatisfaction of internal customers and an absence of information sharing philosophy cross departments. Vertically, overemphasis on the concept of "levels" and "bureaucracy" resulted in an absence of the philosophy of servicing subordinates and team works. Additionally, the inefficient motivation and punishment mechanism as well as an absence of employees' sense of crisis also might produce negative effects.

Innovation revisions. Following the activities of problem diagnoses, managers tried to propose solutions by integrating the original concept of Organization Efficiency Management into the problems existing in the efficiency system of Jiangxi Mobile (i.e., specific problems in process management, performance management, quality management, risk management, organization management and IT supporting system) and their sources (i.e., defects in the operation system, organization structure, human resources and cultural system). On the one hand, as a result, an efficiency management system consisting of six modules was set up, including process management, organization management, performance management, knowledge management, quality management and risk management. In this system, as illustrated in Fig. 3.2, process management serviced as a cut-in point, which put forward the excellence and efficient way through diagnosing the process and finding the gaps between the current efficiency and the long-term goal; organization and performance management serviced as a base, improving the performance of each person and organization and optimizing the processes and ensuring the realization of strategic goals; quality and risk management serviced as a balance, by establishing the risk-early-warning mechanism and quality management rules, which helped the organization to achieve efficient developments and avoid risks at the same time. On the other hand, specific measures for improving efficiency of the operation system, organization structure and human resources and cultural system were put forward respectively, aiming at putting forward specific efficiency improvement measures. (1) More specifically, the efficiency of the operation system could be improved by establishing a customer-centered process system, optimizing key processes, setting up process optimization team covering seven majors (including channel management, value-added business development, customer management, investment management, network operation maintenance and management, financial centralized management and financial audit management), focusing on urgent processes, examining process efficiency by horizontal and vertical crossing of the process optimization, creating a process management system to promote the establishment of a long-term effective mechanism of process management, improving KPI designs by increasing evaluation weights of internal customer satisfaction, and promoting cross-department coordination through strengthening the support to first-line departments from the provincial headquarters. (2) When it came to the aspect of the organization structure, the efficiency of the organization structure could be enhanced by removing horizontal barriers and improving vertical structure designs, and the adjusted organization structure was more market-oriented than the former one, which highlighted the speed of organization response to customer and market, enabled the organization operation more efficient and flexible and eventually promoted the organization efficiency. (3) Moreover, that of the human resources and cultural system could be raised by improving the staff career planning and the system of rewards and punishments, adhering to the core values, mission and vision of the organization to form unified values, keeping the connect between culture management and strategy management, assessing the consistency and fitness between activities of efficiency management and the culture, building up harmonious teams, inspiring enthusiasm for

Fig. 3.2 Efficiency management system of Jiangxi Mobile

entrepreneurship, and constructing a mechanism for innovation, which provided a strong support of business center tasks and core modules of efficiency management.

Proposal evaluation and Selection (the process to accept or refuse). The innovation scheme of efficiency management was multidimensional and complex, referring to not only qualitative problems (such as cultural construction) but also several aspects including process management, organization management, performance management and IT supporting system, and referring to even every corner and every employee of the organization, which made high difficulties to quantify implementation costs and benefits as well as to scientifically assess risks. Consequently, whether to take these solutions into practices mainly depended on evaluations made by top-level managers, especially the general manager. After objectively considering both strengths and weaknesses of these innovative solutions, and predicting their costs and benefits, managers regarded the proposed solutions as an effective method to dramatically improve the efficiency of operation and solve existing problem radically. That is to say, efficiency management practices would be an important weapon for the survival and development of Jiangxi Mobile in the future. At the same time, the risks brought by efficiency management were in the range that the company could undertake and top managers could accept. Therefore, on the basis of the proposal evaluation, combined with the discussion results of top manager meetings, the revised proposal was finally adopted.

3.3.2 Implementation

The adoption decision process was undoubtedly essential, but the value of innovation could not be realized without implementation actions. Hence, implementation is a technical process of establishing value of new management innovation in vivo (i.e., in a real setting) (Birkinshaw et al. 2008). Jiangxi Mobile spent over three years in applying these solutions into practices, which proceeded in three main phases: implementation preparation phase, overall implementation phase and the solidification phase. More specifically, the implementation preparation phase reflected activities of setting up an organizational base on the basis of the allocation of resources, training employees to make the internal diffusion of innovation knowledge, carrying out supporting activities like culture cultivation and knowledge management. In the overall implementation phase, it summed up experience and evaluated outcomes of the previous phase by focusing on key processes optimization, performance improvement and organization improvement. The solidification phase was a process of further improving various approaches and methods of market and internal operation, and transforming efficiency management practices into long-term routines of the company.

Implementation Preparation. During the preparation time, in spite of resources allocation, four main tasks had been taken. Firstly, setting up an organizational base for innovation was an action of building up the Efficiency Management Leader Group, Efficiency Management Office, and Efficiency Improvement Working

Group. Efficiency Management Leader Group consisted of top managers, managers from main functional departments and core professional employees to give directions for implementation; Efficiency Management Office was directly controlled by the strategic development department of the headquarters assisted by related key members of each department, with the responsibilities of studying and making relevant plans as well as keeping all departments and individuals engaging in the implementation process; With the purpose of further specifying designs and implementations of efficiency improvement works and allocating all responsibilities to certain undertakers, Efficiency Improvement Working Group consisted of top managers, department managers, core employees and others from branches. Secondly, in order to stimulate the working enthusiasm of employees and make all staff actively devote to the work of efficiency management, the organization introduced various training projects and activities combined with actual conditions of the organization and psychological characteristics of employees in the implementation process of efficiency management. For example, activities of brainstorming, frank discussion and innovation effect prediction encouraged first-line employees to participate in designs and implementations of the innovation, and to enjoy the feeling of being a master of the organization so as to enable the use of their initiative and creation and eliminate their psychology of hostility. Thirdly, culture cultivation referred to informal discussion meetings and good examples calling that cover various levels of the organization to unify the value of employees', and objectively evaluate the fitness between Efficiency Management practices and organizational culture, which organically combined the publicizing, implementation, internalization and transformation work of culture construction with the key breakthrough of efficiency management, to finally establish an efficient cultural system based on customers satisfaction. Culture cultivation was conducive to the implementation of the activities and systems of efficiency management, and was a support to achieve the operation goal of the organization. Finally, all key breakthroughs of efficiency management needed all-round knowledge supports. Given a lot of problems existing in the knowledge management of the organization and many important factors affecting the overall operating efficiency of it, the organization also attempted to construct a knowledge base and an I-know sharing platform to induce accumulation, diffusion, and sharing of experience and knowledge of certain jobs, which could not only address the problem of valuable knowledge capital losing leaded by job rotation and brain drain but also enable to satisfy employees' demand of pursuing continuous development and gain additional attractiveness and cohesiveness for the organization. This phase of preparation ended in 6 months.

Overall implementation. After making a good preparation and obtaining ideal results for implementation, managers further engaged in activities of systematically improving key operation processes, performance management and organizational structure comprehensively and selectively on the basis of continuing to strengthen cultural cultivation and knowledge management. This phase of overall implementation lasted almost one year.

Especially, in order to improve the efficiency of key operation processes, Jiangxi Mobile first integrated the demand of external customers into internal processes according to the panoramic view of the process based on customers satisfaction and identified key processes that most importantly and urgently needing to be improved from the existing short processes by sticking to its strategy position of becoming a world-class firm. As a result, seven key short processes were identified, including processes of channel management, value-added services development, group customers management, investment plans management, net operation and maintenance, financial centralized management and capital audit management. Then, seven working teams for process optimization were set up, with each in charge of improving one process. For example, the working team for improving the efficiency of value-added business development process in Jiangxi Mobile successfully addressed two core problems of over complexity and unclear division of work and responsibilities among departments, through a sequence of activities referring to identifying key links in the process, resetting up their connections with corresponding departments and redesigning the whole process, as showed in Figs. 3.3 and 3.4. The original process showed that the value-added business development went through a complicated and overlapping process of "new product needs submission (all departments) → evaluation and selection (all departments) → preliminary appraisement (data department) → new product development (product development department) → new product test (data department) → acceptance check (data and product development departments) → commercial trail report (data department) → a preparation for commercial trial (data department and commercial trail unit) → marketing (commercial trial unit) → commercial trial summary report (commercial trial unit) → formal commercialization (marketing department) → products improvement (data department)". On the contrary, the optimized process reflected a much simpler and more efficient sequence of activities involving "new demand search (marketing department) → new business development, test and evaluation, and commercial trial plans making (network department) → commercial trial decision (the headquarters) → commercial trial reports and commercialization plans making (data telecom department) → commercialization decision (the headquarters) → business commercialization and post evaluation(marketing department)".

In order to improve the performance system, Jiangxi Mobile added indexes reflecting reaction speed of key processes to the system; and optimized performance appraisement system based on process synergy; also, raised the weight of internal customer satisfaction; in addition, transformed the existing model where the provincial headquarters were in charge of all performance appraisements to a new model with all its branches and major departments involved in. The new system emphasized the support to the first lines of the organization and the coordination among the background, foreground and various departments in the headquarters. From this perspective, these measures strengthened the support services from provincial headquarters to first-line branches and the collaboration among the background department, the foreground department and each department of the provincial headquarters. Additionally, Jiangxi Mobile improved motivation

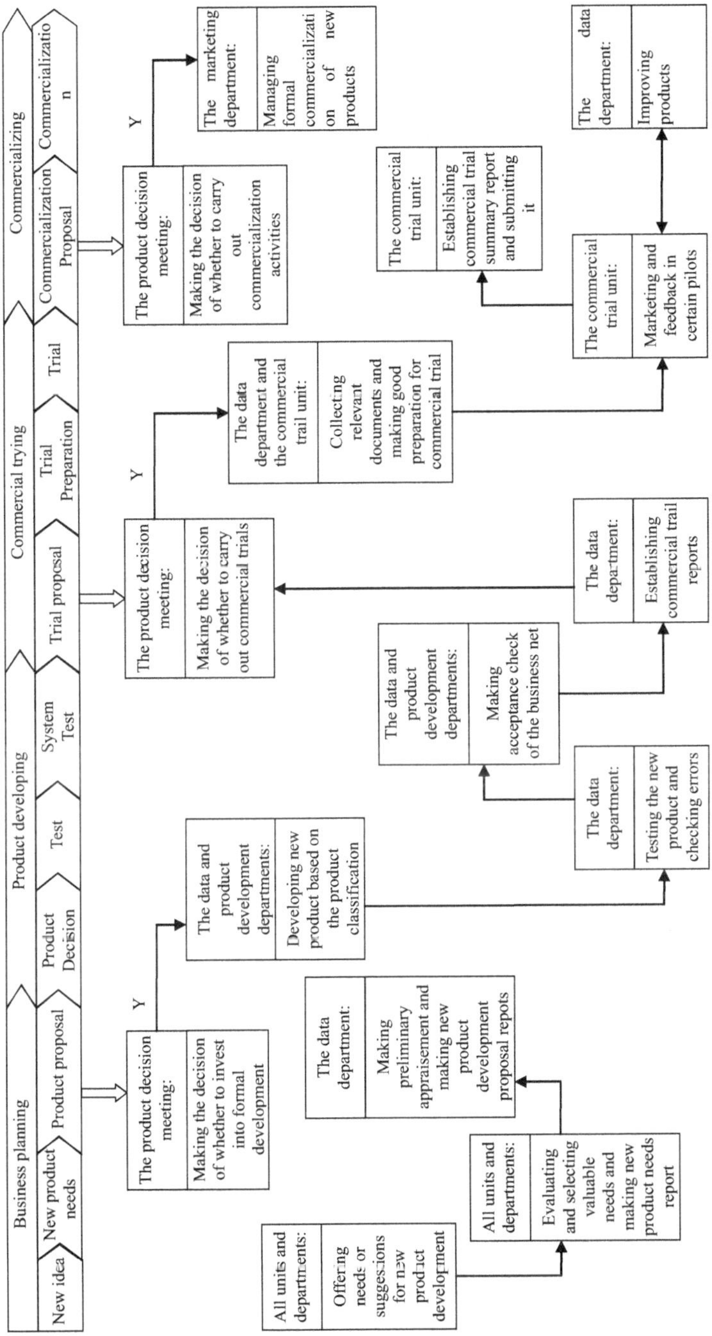

Fig. 3.3 The original process of value-added businesses development

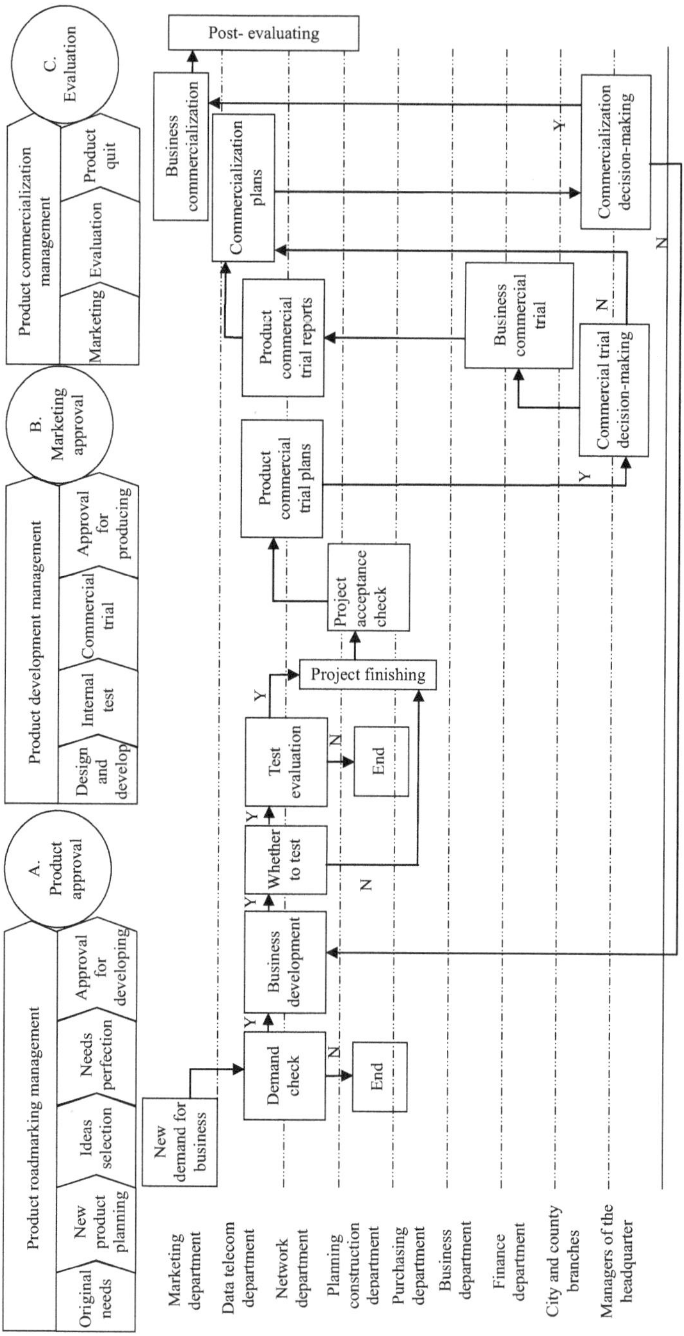

Fig. 3.4 The improved process of add-valued businesses development

mechanism to stimulate working enthusiasm of employees, for example, publishing Guiding Opinions about Employment Integration and Guiding Opinions about Career Development of Professionals, and carrying out experimental units of the special project at the provincial customer call center, with the purpose of offering differentiated career development suggestions for employees in different levels and positions, on the one hand, which resulted in getting through the career development paths. On the other hand, a system of score management was also set up as a complement for the performance appraisement system to induce additional behaviors in accordance with the organizational culture and development as well as to stimulate the spirits of innovation, entrepreneurship and team working from organizations and individuals on the basis of completing the departmental duties and post responsibilities.

In the aspect of improving organization structure, the structure of Jiangxi Mobile was renewed to activate its energy in combination with carding processes. In order to enhance the matching degree of organizational structure to the market development and customer demands, take that of the provincial headquarters of Jiangxi Mobile for example, the organizational structures of which were engaged in an adaptive adjustment, the management center for group customers under marketing department and the industrial customer center of Jiangxi province were replaced by group customer department, as illustrated in Fig. 3.5. Consequently, administration office and department of execution affairs were replaced by a new-established general administration department. Moreover, supervision office and discipline inspection office were merged and renamed office of discipline inspection and supervision. Compared with the original structure, the new one was much more market-oriented, which enabled a higher speed of reacting to requirements of market and higher efficiency of organizational operation and resulted in further improving the organization efficiency.

Solidification. After nearly half a year of preparation and another year of overall implementation, the operation efficiency of Jiangxi Mobile had been dramatically improved with an obvious raise in its market position and competitive competency.

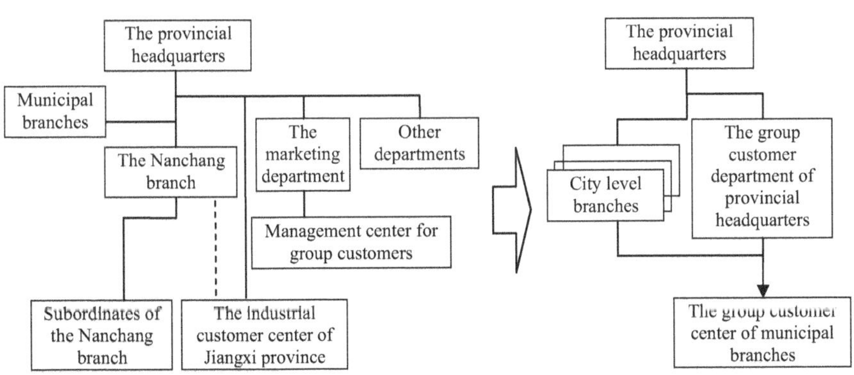

Fig. 3.5 Transformation of Jiangxi Mobile's structure

Then, another two years were spent in chronically solidifying the new practices of Efficiency Management to perform their functions and transforming them into organizational routines after summarizing the experience and effects. Jiangxi Mobile further improved means and tools of Efficiency Management and set up a process management system to ensure long-term functioning of the new practices. This process management system consisted of general planning and management realizing. The general planning referred to activities of defining goals, content, guiding principles, development thought, system framework, life-cycle management, organizational structure, performance appraisement and promotion plans of process management, while management realizing reflected activities of process planning and carding, designing and approving, implementing and promoting, executing and controlling, optimizing and renewing, documents managing, etc. Consequently, Jiangxi Mobile successfully addressed internal problems and met the demands of external environmental changes by adopting and implementing the new practice of Efficiency Management.

In brief, through three links of system diagnostics, identification of problems and their roots, and adaptations between innovation and the problem, Jiangxi Mobile eventually put forward the Efficiency Management Theory System and the specific innovation solutions, and after more than three years of practicing, it had confirmed the effectiveness of innovation solutions. The key point lied in the scientific diagnoses of problems and their roots in the process of adaptations between innovation and the internal problems, and established a close connecting relationship among the innovation solutions, the problems and their roots, enabling innovation to indeed effectively address the internal problems of the organization.

3.4 Discussion

3.4.1 The Framework of Adoptive Management Innovation Process

Building on the exploration into the whole process of Efficiency Management in Jiangxi Mobile and the conception of what makes adoptive management innovation unique, this section assembles the activities of management innovation into two more general but interlinked subprocesses of adoption decision and implementation, as illustrated in Fig. 3.6.

During the adoption decision process, the identification of a novel problem first leads to the dissatisfaction of managers at the operational performance of its organization. In another word, problem identification is considered as the starting point of putting forward and introducing advanced management ideas or methods of an organization, and represents the difference between actual performance and potential performance of the organization (Guillén 1994), while the organization attempts to achieve the goals of adjusting to the environment changes and

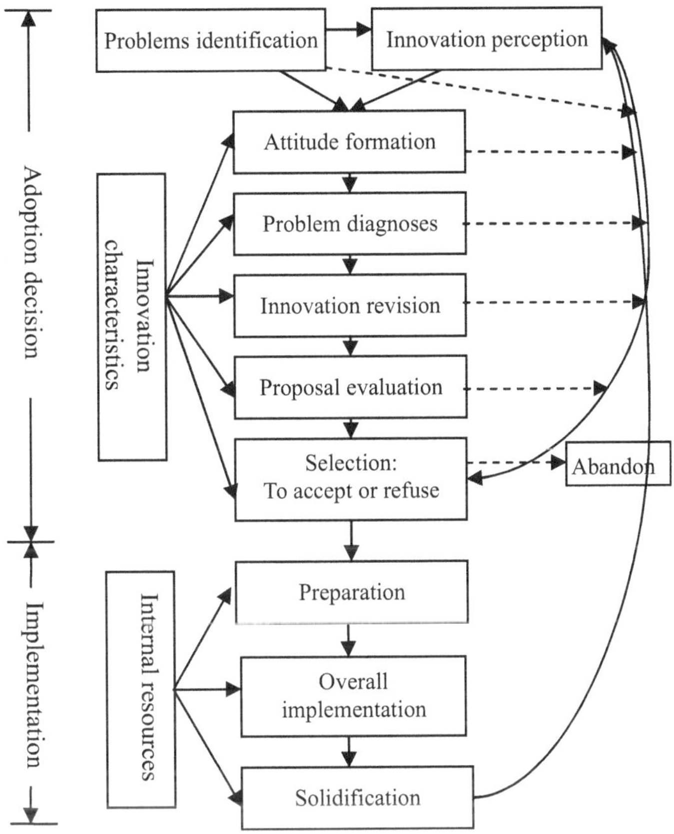

Fig. 3.6 The framework of adoptive management innovation process

improving performance difference through innovations or changes. While, the environment changes mainly refer to the general environment (for example, political, economic, legal environment that can't be controlled by the organization) and the specific environment (for example, competitors, industry development, suppliers and customers). Once perceived environmental changes, the organization has taken active measures to make corresponding changes to adapt the new environment, in recent years for example, with the rise of the Internet and the e-commerce, organizations have been developing innovative activities such as information management and e-commerce construction. Performance perception differences perhaps are caused by actual performance problems or by deviations from expectations of opportunities and environments (Guillén, 1994). Perception differences are the direct expression of dissatisfaction with the status quo, including the low production efficiency, the dissatisfaction from employees of compensation system, performance appraisal system and employment system, problems that are difficult to coordinate between different departments, and the crises and challenges, and running problems, etc. Similarly, once perceived organizational performance differences, top managers will have the idea of changing organization management and operation mechanism.

In fact, there are some differences between different categories of management innovation. From the four aspects of measuring the introduction of new management practices proposed by Mol and Birkinshaw (2009): (1) whether to implement the new strategy, (2) whether to implement the advanced management technology, (3) whether to implement new organizational structures or to make significant changes to the organizational structure, (4) whether to change the organization marketing strategy, and referred to the classification of the management innovation in the recent "National Enterprise Management Innovation Achievement Award", this chapter will divide management innovation into nine categories, consisting of strategic change, management technology innovation, organization structure innovation, production and operation process innovation, organizational culture innovation, innovation of marketing ideas and operation channels, risk control and financial management innovation, human resource management and performance management innovation and other management innovation. The specific reasons for these nine innovations are different, as shown in Table 3.5.

Table 3.5 Management innovation categories and reasons

Management innovation categories	Management innovation reasons
Make significant changes to existing strategy by introducing new strategic management ideas	External reasons: increasingly complex environmental pressure, dynamic competition pressure Internal reasons: the change of enterprise resources and enterprise organization system, the delay of the original enterprise strategy, the restriction of complex enterprise system, the upper limit of enterprise's growth, etc. (Jie 2006)
Introduce more advanced management techniques, for example, knowledge management, quality management, etc.	Outdated management techniques, low organizational efficiency, the promoted requirements of organizational performance and overall external competitiveness, assists in developing core technical abilities, etc.
Introduce or change organizational structure, such as implementing the flat organization structure	To flexibly adapt to dynamic changes of external environment, to meet the development needs of organizational flexibility, to assist in the implementation of elastic strategy, to improve internal communication efficiency, to reduce management costs and to improve management efficiency, etc.

(continued)

Table 3.5 (continued)

Management innovation categories	Management innovation reasons
Adjust the production or operation process of the company, such as introducing and implementing business process reengineering and refined production, etc.	Too many processes or repetitive running programs; the needs of information development; the improvement of process efficiency and quality; the implementation of process management; to be consistent with the development of other departments, etc.
Create organizational culture, such as building a learning organization, etc.	The outdated concepts of employees, a lack of creating organization culture and values, a lack of softwares supporting operation and development, etc.
Introduce new marketing ideas or channel operations such as green marketing	The outdated marketing concepts or channel operation modes, low efficiency, the needs of dynamic marketing changes, the changes of social concepts and consumption concepts, etc.
Introduce new risk control or financial management methods	A lack of efficient risk control methods, the lagging of existing financial management systems and methods, the needs of resisting risks, to realize values of organizational resources, etc.
Introduce new human resource management or performance management methods or systems	To optimize the existing human resource management or performance management methods or systems, to meet the demands of employees, to abstract more excellent individuals, etc.
Other innovation, for example, intellectual property management	To enhance the market competitiveness and to strengthen the development capacity of the organization

Although the reason or purpose of carrying out all kinds of management innovation varies from each other, it still exists something in common, namely, most innovation are made to solve the relatively abstract and qualitative problems of the organization, which are not easy to find out directly by data statistics or various statements for managers. Therefore, it is necessary for top managers to carry out subjective perception activities in practice to identify problems. In addition to the effects of entrepreneurial experience and knowledge base, the problem perception or identification mainly depends on the innovation intention of entrepreneurs, that is, entrepreneurs with strong innovation intention are more sensitive to environmental changes and performance differences and are more likely to produce dissatisfaction to organization status quo, enabling entrepreneurs a stronger tend to take actions to change the status quo, thus implementing management innovation. At this moment, managers do not find the source of existing problems, but only make an initial judgment according to various phenomena.

Driven by the new problem identified and dissatisfaction of managers, an organization comes up with problem-driven search to obtain an existing management practice through intended activities (Williams and Rao 1998), recognizes an opportunity of using it (Wilson 1987; Zaltman et al. 1973), and seeks further information to gain a better understanding of it, for example, collecting ideas from employees, learning experience of other organizations, and obtaining mature ideas from management ideologist, scholars, management heroes and consultants. Mol and Birkinshaw (2009) propose that, establishing more knowledge source connections or establishing connections with more types of knowledge sources can positively affect the introduction and implementation of new management practices. The technical innovation literatures emphasize the driving effect of various external knowledge sources such as leading customers and suppliers on innovation. The more the knowledge sources are, the higher the frequency of knowledge extraction is, the more comprehensive and profound the entrepreneur's understanding of the new management practice is, and thus the higher the success rate of implementation is. The more diverse the knowledge sources are, the higher the likelihood of innovation and value reintegration of obtained inspirations is (Hargadon 2002). Mol and Birkinshaw (2009) also suggest that the broader the knowledge scope is, the higher the level of introducing management practices is. Simultaneously, in addition to the knowledge sources, managers also attempt to perceive the comparative advantages, compatibility, complexity and observability of the new practice (Hashem and Tann 2007). To perceive innovation characteristics not only influences the formation of attitude, but also influences the final decisions (Hashem and Tann 2007). Combining the problem identified and innovation perceived, managers form an attitude towards the innovation (Williams and Rao 1998; Zaltman et al. 1973). A positive attitude leads to positive preference and intention to further involve in, and in turn induces the organization to initiate innovation (Williams and Rao 1998). Conversely, a negative attitude may make an organization exit from the process. However, a formed positive attitude or the initial innovation ideas are not enough to enable the managers to make final decisions. Once a positive attitude formed, it's time to explore into the sources of the problems identified and reestablish a new scheme combined with the acquired innovation knowledge, which indicates a sign of the difference from technology innovation adoption. Although in the initiation phase, a positive attitude toward certain existing practice has been formed, it does not enable managers to introduce the new practice directly from another context because any organization is a unique, complex organic body. That is, the practice finally implemented is not exactly the existing one, but a new scheme revised by integrating the original idea into organizational environment and resources. For example, after identifying the problems of low operational efficiency at the beginning of initiation phase and forming a positive attitude toward Organizational Efficiency Management, Jiangxi Mobile investigated the roots of the problems deeply and characteristics of its own, and then set up a new specific plan only fitting for itself. From a theoretical perspective, it still belongs to the Efficiency Management, and the specific measures that adapted to the organizational context are different from that or any other organization practices in the initiation stage.

Following the activity of innovation revisions, the organization begins to objectively and scientifically evaluate the expected risks and benefits of the new plan by considering its characteristics, technological demands, costs and outcomes to decide whether to accept it or not (Williams and Rao 1998). Management innovation may potentially require fundamental changes in the routines or DNA of organizations, which makes it too difficult to undertake in an effective manner, and significantly harder than the process of technological innovation where the innovation is relatively more tangible and less systematically dependent, in addition, most organizations lack existing expertise or experience of management innovation (Birkinshaw et al. 2008), and management innovation has characteristics of ambiguity, complexity, and result-uncertainty, it's really hard for managers to evaluate that with scientific measures. Therefore, managers would not objectively evaluate innovations but subjectively perceive potential risks and benefits. Finally, yes-no selection will come naturally. Managers may refuse it when their perceived risk is higher than benefit, namely ending the management innovation process; while, adopt it and transit to implementation phase when their perceived risk is lower and acceptable. At that point, managers will combine the internal and external environmental changes, personal experience, and other aspects of factors, at the same time, the feedback information provided from all stages of the decision-making process may be a reason for an organization to give up or introduce the innovation (Hashem and Tann 2007). Risk perception is an important aspect of the differences among managers and belongs to the process of individual cognition, which has obvious subjectivity. Therefore, the differences between the perceived and the actual risks exist inevitably: if the perceived risk is higher than the actual one, then refuse the innovation and lose the opportunity to create competitive advantages for the organization; while if the perceived risk is lower, then to introduce the innovation may become a risky behavior which is hard to achieve the expected effects. Visibly, when managers make the final decisions, it is necessary to put more emphasis on opinions of other management team members or subordinates, as well as success or failure cases from other organizations, and finally avoid the serious consequences of the blind impulsive decision.

Unless an organization exits from the innovation, the implementation process begins after the adoption decision. Implementation is the process of implementing the innovation values and has a stronger operability. The case analysis results show that, implementation, which relies much on the internal resources of an organization, consists of events and actions that pertain to modifying the innovation, preparing the organization for its use, trial use, and acceptance of the innovation by the users and continued use of the innovation until it becomes a routine of the organization (Damanpour and Schneider 2006; Duncan 1976; Meyer and Goes 1988; Rogers 1995). In this phase, the innovation is put into use by organization members. This research proposes a three-phase process for implementation of adoptive management innovation. The three phases, i.e., preparation, overall implementation and solidification, are indispensably interrelated, which is consistent with existing research on management innovation processes. As illustrated in Fig. 3.6, the preparation phase is a process of making good preparation for adopting a new management practice, for example, establishing the basic work of organization guarantee, resource allocation and staff training, aiming at reducing the risk of comprehensively implementing the scheme; the overall implementation phase is concerned with actions of determining the scope of implementation and the areas or objectives implemented emphatically and transforming the proposed innovation plan into practices and producing outcomes after summing and solidifying results of the previous stage; and the solidification phase is a process of transforming new practices into organizational routines which would bring long-term effects on organization performance, the main task of this phase is to strengthen or maintain innovation practices that gained positive effects and finally achieve the purpose of bringing continuous competitive advantages to the organization.

3.4.2 Implications for Theory

The introduction and implementation of management innovation is a huge system project, and its process has a significant complexity. This chapter focuses on the nature of adoptive management innovation, and its two-subprocess framework through an exploratory case study on Organizational Efficiency Management innovation of Jiangxi Mobile in China. It offers various implications to existing literature on management innovation by addressing the question of how adoptive management innovation occurs.

This chapter reflects the argument made by Birkinshaw and Mol (2006), Birkinshaw et al. (2007, 2008) that the generative mechanisms of management innovation through which it occurs are theoretically interesting in their own right and also relatively poorly understood. Birkinshaw et al. (2008) develop a framework based on new-to-the-state-of-the-art management innovation highlighting four interlinked phases of motivation, invention, implementation and theorization and labeling, while this research proposes a general process with two key interlinked subprocesses of adoption decision and implementation for adoptive management innovation, each containing various activities. Nevertheless, there are some common findings. For example, both of them stress new problem identification in the early period of innovation. Birkinshaw et al. (2008) argue that the demand for new management practices is driven by the identification of a novel problem—a perceived shortfall between the organization's current and potential performance (Barley and Kunda 1992; Guillén 1994). Similarly, this research argues that this perceived shortfall or problem identification directly leads to the dissatisfaction of managers at low operational efficiency, poor coordination among departments, new crisis and challenges faced with and so on, and further promotes the emergence of management innovation. That is to say, the purpose of introducing management innovation is to solve the problems of the organization. Additionally, both require a complex and multi-stage process and reveal that implementation, through which the value of the new management practice could be realized, is an indispensable phase in management innovation. In addition to the implementation stage, the innovation process also needs to carry out a large number of early-stage decision-making activities.

In comparison with the research of Birkinshaw et al. (Birkinshaw, J., Mol, M.J.: How management innovation happens. Mit Sloan Manag. Rev. 47(4), 81–88 2006; Birkinshaw et al. 2007, 2008), this research on new-to-the-organization or adoptive management innovation differs on four aspects at least, which indicates to some degree the differentiation between the two types of management innovation. Firstly, Birkinshaw et al. (2008) see invention of a new hypothetical management practice as the phase following problem identification, while this research stresses outside perception of existing management practices. This indicates different innovation sources: generating innovation is created by internal agents in response to a specific problem or opportunity, focusing on the problems of innovation ideas of "invention" or the new knowledge creation; and adoptive innovation is introduced by external agents from somewhere else, with the emphasis on the acquisition of innovation knowledge. Secondly, instead of the activity of trials and errors in which progress is achieved by monitoring and making adjustments against the original concept mentioned by Birkinshaw et al. (2008), this framework emphasizes activities of problem diagnoses and realization of the fitness between introduced management practices and the new organizational context. That is, introduction of existing management practices or methods from the external is not only a simple process of knowledge transferring, but a more complicated and logical process through which the existing practice perfectly fits to the new organizational context and new schemes are proposed, which is embodied as a process from external to internal; while The proposal of the all-new innovation scheme is based on the organization problems, so it is more adaptive and reflected as a process from internal to external.. Thirdly, the framework on generative mechanisms of completely new innovation stresses the important roles of both internal change agents (mainly including managers or entrepreneurs) and external change agents(mainly containing external consultants and management elites, etc.) in the process as well as the ways these two sets of actors interact with one another (Birkinshaw et al. 2008), believes that employees are the fundamental elements or initiators of management innovation, so the completely new innovation is a process of all-staff participation, bottom-up and external support; while this research, mainly putting attention on the adoptive management innovation which is a top-down process, focuses on internal agents only, especially the roles of core managers who almost dominate the progress of the whole process in adoption decision. According to the practice of Chinese firms, it has been found that the phenomenon of over reliance on internal agents is rooted in the prevailed family business management pattern where the owner or the general manger holds most power for decision making. Furthermore, underdevelopment of the management consulting industry in China may also lead to an absence of support from the external. Finally, theorization and labeling, the fourth phase in the framework of generating innovation, was not mentioned in this research. Though studies have shown a significant effect of theorization and labeling on the acceptability of management practices to various constituencies (Eccles and Nohria 1992; Kieser 1997; Birkinshaw et al. 2008), this research finds that these activities have not attracted attention of firms in China. The purpose of theorization and labeling is to apply new ideas and achieve initiative acceptance and participation of all staff. Most internal agents subjectively think it is unnecessary to take these activities because adoptive innovation has been theorized and labeled by other organizations where they are created and implemented successfully. More importantly, under the order-dominated and high-centralization management pattern, though employees always have to accept innovation concepts and participate in innovation practices passively, they seldom present their unwillingness or make any resistance, which covers the importance of theorization and labeling or re-theorization and re-labeling. Additionally, in the implementation process of adoptive management innovation, due to the existing successful experience, the managers are more confident, and the employees' resistance is relatively lower, making the implementation process smoother. However, because of the higher degree of uncertainty and risk, the generating innovation is easier to lead to employees' resistance that undermines the implementation of innovation. It is worth noting that this also tends to cause overconfidence and cognitive bias of managers, which further underestimates the risk of adoptive innovation and leads to the high failure rate of innovation.

3.4.3 Implications for Practice

Adopting a longitudinal interpretive and exploratory case study and blending innovation process and internal agent literature enabled development of a process framework that explicated realization mechanism of adoptive management innovation on the context of China. More specifically, this work identifies phases of innovation, illustrates roles of internal agents in each phase and theorizes the potential for realizing the systematical project of change. This framework suggests a number of important insights for practice.

Though adoptive management innovation indicates introduction and implementation of management practices implemented successfully somewhere else, it is not a simple job of repetition, but still a huge project with a complicated process full of uncertainty. When joining the wave of management innovation, a large number of Chinese firms, especially middle and small ones, underestimate the risk of adoptive management innovation and take actions blindly. They realize the potential of adoptive innovation in improving management and even performance of the whole firm, but neglect the accompanying risk and uncertainty. Therefore, this research shows how adoptive management innovation comes about to remind firms of risks and difficulties.

In this research, the established framework shows differentiation between two types of management innovation and uncovers the key for the success of adoptive innovation. It offers some implications for the practice directly. When making a choice between generative and adoptive innovation, firms need to consider about their differentiated nature or definition; and more importantly, pay attention to different processes of realization. It confirms that the success of adoptive management innovation relies on their adaptation to their idiosyncratic context within the organization where they are adopted, as advocated by Ansari et al. (2010) and Vaccaro et al. (2012). Especially, the findings show that problem diagnoses and innovation revisions are two key phases in the process of adoptive management innovation. Accordingly, the reasons for the failure of numerous organizations might be a lack of these key phases in introduction and implementation of new practices like Business Process Reengineering, Customer Relationship Management and Total Quality Management. Moreover, the sequence of activities in the whole process is also a core element in this research. For example, managers of an organization with too much thought may have several initiatives that are well progressed in adoption decision, but with no commensurate investment in implementation. In another case, an organization may see implementation as a means of establishment where initiatives are worth pursuing, with the process of adoption decision ignored.

3.5 Conclusion

Management innovation is not achieved overnight, but a gradual process. This chapter addresses how management practices successfully implemented somewhere else, namely adoptive management innovation, has been introduced into a new firm and then effectively implemented, or investigates key activities of the generative mechanisms through which an adoptive management innovation occurs. Based on existing literatures on management innovation and the nature of adoptive management innovation, it sets up a two-subprocess framework of adoptive management innovation by adopting an exploratory case study on Organizational Efficiency Management innovation of Jiangxi Mobile in China. The one subprocess, adoption decision, mainly consists of themes of problem identification, innovation perception, attitude formation, problem diagnoses, innovation revisions, proposal evaluation and selection. The other subprocess, implementation, has three main phases: implementation preparation phase, overall implementation phase and the solidification phase. The findings offer many valuable insights for further research in how to explore the generative mechanism of changes or innovations in special context of China and hold important implications for management practices. It broadens Rogers' (1995) research on innovation adoption by focusing on management innovation, but not technological innovation, and stressing the role of managers and the process of innovation implementation. Therefore, in the practice of introducing management innovation, the managers need to pay special attention to the adaptation between the existing innovation and organizational conditions on the basis of reviewing the practices of other organizations. In other words, though the implemented scheme stems from the outside, it must undergo the internal processes of problem diagnoses and scheme proposing, only in this way can it avoid making a lot of firms suffer heavy losses rather than achieve the desired effects after introducing business process reengineering. Additionally, it has presented the internal generative mechanism of adoptive management innovation.

References

Abrahamson, E.: Management fashion. Acad. Manag. Rev. **21**(1), 254–285 (1996)

Andriopoulos, C., Lewis, M.W.: Exploitation-exploration tensions and organizational ambidexterity: managing paradoxes of innovation. Organ. Sci. **20**(4), 696–717 (2009)

Ansari, S.M., Fiss, P.C., Zajac, E.J.: Made to fit: how practices vary as they diffuse. Acad. Manag. Rev. **35**(1), 67–92 (2010)

Barley, S.R., Kunda, G.: Design and devotion: surges of rational and normative ideologies of control in managerial discourse. Adm. Sci. Q. **37**(3), 363–399 (1992)

Birkinshaw, J., Crainer, S., Mol, M.: Special report on management innovation. Bus. Strategy Rev. **18**(1), 45–73 (2007)

Birkinshaw, J., Hamel, G., Mol, M.J.: Management innovation. Acad. Manag. Rev. **33**(4) (2008)

Birkinshaw, J., Mol, M.J.: How management innovation happens. Mit Sloan Manag. Rev. **47**(4), 81–88 (2006)

Damanpour, F., Schneider, M.: Phases of the adoption of innovation in organizations: effects of environment, organization and top managers. Br. J. Manag. **17**(3), 215–236 (2006)

Duncan, R.B.: The ambidextrous organization: designing dual structures for innovation. In: Kilman R.H., et al. (eds.) The Management of Organization: Strategy and Implementation, vol. 1. pp. 167–188 (1976)

Eccles, R.G., Nohria, N.: Beyond the Hype: Rediscovering the Essence of Management. Harvard Business School Press, Boston (1992)

Eisenhardt, K.M.: Building theories from case study research. Acad. Manag. Rev. **14**(4), 532–550 (1989)

Guillén, M.F.: Models of Management: Work, Authority, and Organization in a Comparative Perspective. University of Chicago Press, Chicago (1994)

Gummesson, E.: Qualitative Research Methods in Management Research. Sage, Thousand Oaks, CA (2000)

Hamel, G.: The why, what, and how of management innovation. Harvard Bus. Rev. **84**(2), 72–84 (2006)

Hargadon, A.B.: Brokering knowledge: linking learning and innovation. Res. Organ. Behav. **24** (02), 41–85 (2002)

Hashem, G., Tann, J.: The adoption of ISO 9000 standards within the Egyptian context: a diffusion of innovation approach. Total Qual. Manag. Bus. Excellence **18**(6), 631–652 (2007)

Jie, X.W., Zhang, L.M.: Entrepreneurship Strategy Management. Tsinghua University Press (2006)

Kiesler, A.: Rhetoric and myth in management fashion. Organization **4**(1), 49–74 (1997)

Lincoln, Y.S., Guba, E.G.: Naturalistic inquiry Sage. Naturalistic inquiry. Sage Publications, Thousand Oaks, CA (1985)

Meyer, A.D., Goes, J.B.: Organizational assimilation of innovations: a multilevel contextual analysis. Acad. Manag. J. **31**(4), 897–923 (1988)

Mol, M.J., Birkinshaw, J.: The sources of management innovation: when firms introduce new management practices. J. Bus. Res. **62**(12), 1269–1280 (2009)

Rogers, E.M.: Diffusion of Innovations, 4th edn. The Free Press, New York (1995)

Strauss, A., Corbin, J.M.: Basics of Qualitative Research: Grounded Theory Procedures and Techniques. Sage, Newbury Park, CA (1990)

Su, J.Q., Lin, H.F.: Research on the process mechanism of introduced management innovation. Sci. Sci. Manag. S&T **31**(1), 34–41 (2010)

Vaccaro, I.G., Jansen, J.J.P., Bosch, F.A.J.V.D., Volberda, H.W.: Management innovation and leadership: the moderating role of organizational size. J. Manag. Stud. **49**(1), 28–51 (2012)

Walsham, G.: Interpretive case studies in IS research: nature and method. Eur. J. Inf. Syst. **4**, 74–81 (1995)

Williams, L., Rao, K.: Information technology adoption: using classical adoption models to predict AEI software implementation. J. Bus. Logistics (1998)

Wilson, D.T.: Merging adoption process and organizational buying models. Adv. Consum. Res. **14** (1), 323–325 (1987)
Wu, H.F., Yu, Y.D.: A shallow discussion of the case study method in management: characteristics, method design and effectiveness discussion. Science **22**(S1), 105–111 (2004)
Yin, R.K.: Case Study Research: Design and Methods, 2nd edn. Sage, Newbury Park, CA (1994)
Zaltman, G., Duncun, R., Holbek, J.: Innovations and Organizations. Wiley, New York (1973)

Chapter 4
How Do Managers Make the Decision of Adoptinga Management Innovation?

4.1 Introduction

In recent years, as the market increasingly homogenized, competitive advantages resulted from technology, business or service differentiation tend to be weakened, while management innovation has become a new competitive advantage by promoting the organization efficiency or performance. Its importance is increasingly highlighted. Hamel (2006) emphasizes that, under today's business competition, management innovation ensures the healthy and stable development of organizations by fundamentally addressing organization operation problems, updating organization structures and adjusting organization development direction. Compared with foreign countries, through management innovation was introduced by the domestic enterprises in recent years, it has attracted more and more attentions. Managers gradually realize the gaps between domestic enterprises and the international advanced enterprises in operational efficiency, management and operation modes, etc. Therefore, firms start to develop all kinds of management innovation activities, such as Haier's "Market Chain" Business Process and Alibaba's consumer-to-consumer portal Taobao, which cause a wave of management innovation and gradually establish the important status of the management innovation all over China. There indeed exist some successful achievements global companies have made in the aspect of management innovation over the past century, such as "M-form" of GM, Toyota Production System and Balanced Score Card proposed by Kaplan, which is new to the state of the art without known precedent and extremely rare. More enterprises introduce the existing management practices, namely adoptive management innovation, to a new organization or field. However, existing management practices need to be effectively embedded when introducing to a new organization, which means to rebuild innovative solutions combining with the problems and resources of the new organization, therefore, the implementation of adoptive management innovation undoubtedly is a complex process.

As a big issue concerning the overall benefit of organizations as well as their future development, whether to introduce a new management practice or method

H. Lin, *Adoptive Management Innovation*,
https://doi.org/10.1007/978-981-10-7712-8_4

depends on core managers (Elenkov et al. 2005; Kimberly and Evanisko 1981) who take the responsibility of directing future development of organizations by making essential decisions. Why organizations in the same industry or surrounded by similar environment seldom take the same actions of management innovation? The difference of managers in decision making concerns a lot. However, what actions do managers take during the process of making decisions? And, what are the causes for the differentiation? In order to address these questions, this chapter attempts to establish a three-dimension decision model based on an further exploration on decision process of adoptive management innovation and responsibilities taken by core managers in each stage, and to further examine the set-up model by collecting data from 237 managers, with the purpose of uncovering affecting paths and internal mechanism of complex decision making.

4.2 Adoption Decision Process and the Roles of Managers

Based on the conception of what makes adoptive management innovation unique, the previous chapter has attempted to investigate the case of the Organizational Efficiency Management of Jiangxi Mobile in China to identify key activities of adoptive management innovation and develop a two-interlinked-subprocess framework of adoption decision and implementation. This chapter furthers the research by focusing on the previous decision-making process of adopting new management practices to outstand the key roles of core managers; while the next chapter will focus on the implementation process of innovation. Adoption decision reflects the initiation phases of problem identification, innovation perception, attitude formation, problem diagnoses, innovation revisions, proposal evaluation and yes-no selection. When setting up the whole-process framework, this research focuses on the activities taken by core managers, with the purpose of extracting affecting mechanism of managers on introducing new practices.

In order to reset up the process framework of innovation decision, a multiple case study on a set up adoptive management innovation was conducted. Eight case samples were selected from the National Enterprise Management Modernization Innovation Achievements and the China Academy of Management Awards, including Knowledge Management of China Mobile group Fujian Co., Ltd. (Fujian Mobile), Standardization Production of Shanghai Volkswagen Co. Ltd. (Shanghai Volkswagen), Computer Integrated Manufacturing of China North Vehicle Group Dalian Locomotive & Rolling Stock Co., Ltd. (Dalian Locomotive & Rolling), Brand Construction of Dragon Group Holdings Ltd. (Dragon Group), "Market Chain" Business Process Reengineering of Haier Group Co., Ltd. (Haier Group), Special Product Line Profit Center Construction of Zhejiang Geely Holding Group Co., Ltd. (Geely Group), Precise Management of Inner Mongolia Yili Industrial Group Co., Ltd. (Yili Group), and Low-cost Carriers Business Model of Spring Airlines Co., Ltd. (Spring Airlines), all of which had the principles of typicality, sample focus, rich information and the largest variability, as listed in Table 4.1.

Table 4.1 Sample firms and their management innovation cases

No	Firms	Industries	Ownership	Scale	Location	Innovations	Description of management innovation
1	Fujian Mobile	Telecom	State owned	Large	Fujian Province	Knowledge Management	To transform the implicit knowledge hiding in employees and lower level braches to shared explicit knowledge, and to evaluate, renew and spread knowledge
2	Shanghai Volkswagen	Motor	Joint venture	Large	Shanghai	Standardization Production	To set up standards for manufacturing each module through modularization management, and to make sure all set-up standards performed and continuously improved based on institutions and culture construction
3	Dalian Locomotive and Rolling	Locomotive & rolling	State owned	Large	Liaoning Province	Computer Integrated Manufacturing	To improve computer-aided design level, optimize the product structure design, accelerate the pace of new product development, expand and improve the network, and implement engineering data management system
4	Dragon Group	Clothing	Private	Middle and small	Zhejiang Province	Brand Construction	To strength product quality controlling and protection of intellectual property right, and to implement new brand marketing strategies, based on differentiate brand positioning and brand culture
5	Haier Group	Appliance	State owned	Large	Shandong Province	"Market Chain" Business Process Reengineering	To link every employee's work with the market, which can be an external or internal market, with employee compensation tied to market performance, every employee provides the best performance to meet his or her customers' needs
6	Geely Group	Motor	Private	Large	Zhejiang Province	Special Product Line Profit Center Construction	To set up a profit center for each product from development, to purchase, manufacture and selling, and to implement management by objective
7	Yili Group	Dairy	State owned	Large	Inner Mongolia	Precise Management	To implement quantitative management, stress efficiency and procedure, and realize continuous development by surrounding demands of customers
8	Spring Airlines	Air transportation	Private	Middle and small	Shanghai	Low-cost Carriers Business Model	To add low-cost businesses, and lower down costs to make tickets prices lower that other airlines

They are typical innovation cases from typical firms, industries and context, covering industries of telecom, motor, locomotive & rolling, clothing, appliance, dairy and air transportation, including state-owned, joint venture and private ownerships, belonging to large or middle and small firms, and locating at different areas of China.

Based on the fundamental descriptions of these sample firms and their innovation cases from the National Enterprise Management Modernization Innovation Achievements and the China Academy of Management Awards, a series of deep investigations were conducted to get more data. Multiple approaches were used during data collection to meet criteria for trustworthiness (Lincoln and Guba 1985; Yin 1994), including semi-structured interviews, archival data, and observation. Among them, semi-structured interviews were the major approach. Specifically, a number of core managers in charge of these innovations or made innovation decisions were interviewed. Before each interview, a list of questions was proposed and discussed in the research group. The interviews focused on two aspects: the first aspect was about basic information on sample firms and innovation cases, such as, the development process of firms, their products and performance, operation state quo, innovation background, innovation content, and innovation antecedents; the second aspect referred to why managers made the innovation decision and how, i.e., the considering process of adopting a new management practice. The interviews lasted one to three hours. Interviews always began with questions covering more general topics. In view of the inductive aims, managers were encouraged to go deeply into the details of making innovation decisions. Based on the process framework set up in the previous chapter, managers were asked to answer deeper questions about different phases of making innovation decisions. In addition to listening, the author still went into their ideas, understood their values, motivations, logics and other factors controlling their actions. Moreover, in order to avoid the phenomena that information was not complete or the problems distorted, the author also collected archival data in forms of the innovation proposal, relevant meeting notes, newsletters, memos, annual reports and innovation performance reports, and gathered various articles, media reports, stories and Web materials related to these firms and innovation cases to support this research.

The multiple-case study confirms that, it is core managers who launch an innovation through problem identification, get innovation knowledge for the identified problem and form a positive innovation attitude. For example, in the case of Special Product Line Profit Center Construction in Geely Group, the chief financial officer Yin Daqing first identified the problem of low efficiency in operation system and then tried to get knowledge or information on innovation to address the problem. Yin Daqing got innovation knowledge on the innovation of Special Product Line Profit Center Construction through a period of working in Du Pont USA where the new management practice of Profit Center Model was created and successfully implemented. Considering the connection between the identified problem and innovation knowledge on Profit Center Model, he formed an attitude toward innovation adoption. Similarly, in the case of Brand Construction in Dragon Group, Wang Peihuo, its Chief Executive Officer, also the chairman of the board, realized the importance of brand construction in clothing industry and the disadvantage in competition especially in the international market with a lack of a brand. He visited a lot of clothing firms with famous brands in Italy and Germany with his management team, participated in all kinds of international fashions shows and fashions festivals, and read fashions magazines to get information on clothing brands. Moreover, he established a cooperation relationship with Lamy Brand Consultant, Italy, and constructed a new brand with Chinese cultural characteristics together.

However, the case study also shows that, in the phase of problem diagnoses and innovation revisions, core managers prefer to deliver responsibilities to lower-level managers or employees. For example, in the case of Knowledge Management in Fujian Mobile, the general manager asked department managers to find out existing problems and revising the set-up knowledge management proposals accordingly; in the case of Precise Management in Yili Group, the general manager set up a team for innovation and asked the members to collect specific information on operation procedures and demands of customers, and reset up an innovation proposal by combing theories and principles of precise management with existing problems.

Therefore, during the process of making innovation adoption decision, core managers of a firm are directly responsible for problem identification, innovation perception, attitude formation, proposal evaluation and yes-no selection, but not activities of problem diagnoses and innovation revisions.

Further, in addition to the factors such as management experience and the knowledge base, problem identification mainly depends on managers' innovation intention. Therefore, problem identification reflects the innovation intentions of a manager. As Krueger and Carsrud (1993) defines, innovation intention reflects a psychological feature or willingness of investing a great deal of human and material resources in adopting and implementing a new management practice, process, structure, or technique to improve operative performance. It is always used to describe the preference of managers to innovations, and adopted as a best predicting indicator for innovative behaviors. Strong intention drives managers to frequently scan and perceive dynamics of external environment and the shortfall between the organization's current and potential performance (Barley and Kunda 1992; Cyert and March 1963; Guillén 1994). A perceived shortfall, which can be caused by a problem that undermines current performance or by opportunities that may exist and the anticipation of environmental changes (Cyert and March 1963; Ocasio 1997), leads to the dissatisfaction of individual managers who then initiate an innovation to change the state of quo. In fact, the process of perceiving environmental dynamics and shortfall is also that of identifying a novel organizational problem which may drive the demand for new management practices.

Innovation perception is based on innovation knowledge acquirement of a manager. Huber (1991) defines knowledge acquirement as a process through which knowledge is reoccupied by another individual or organization. It highlights the obtaining or grasping of new knowledge. Driven by strong innovation intention, managers come up with problem-driven search to obtain relevant knowledge through intended activities, or to identify available opportunities (Williams and Rao

1998; Zaltman et al. 1973). Mol and Birkinshaw (2009) point out that the introduction of management innovation is essentially a process that managers search, acquire and effectively combine with the enterprise knowledge and resources through new management knowledge. In the process of acquiring innovation knowledge, managers percept and evaluate the advantages and characteristics of innovation itself through subjective consciousness (Hashem and Tann 2007). Based on the acquired innovation knowledge, managers attempt to perceive the comparative advantage, compatibility, complexity and observability of new practices (Hashem and Tann 2007) and form a positive or negative attitude to them.

Moreover, the stage of yes-no selection will come naturally when the previous four are finished. Managers tend to refuse it when perceived risk is higher than benefit, and adopt it when perceived risk is lower and acceptable. However, a gap between perceived and real risk does always exist: if the perceived risk is higher than the actual one, there is no doubt that managers choose to give up innovations, which means losing a great opportunity to create competitive advantage for the organization; on the contrary, if the actual risk is higher, that managers make a decision to introduce the innovation may become a risky behavior, which probably cannot get the expected effect.

All in all, in the process of management innovation introduction and decision, the core managers play an irreplaceable role. Namely, core managers with stronger innovation intention will actively identify the change of internal and external environment and produce innovation intentions, then obtain existing innovation practices or methods from external environment. After delegating problem diagnoses and scheme making tasks, they will evaluate the scheme through risk or benefit perception to finally decide whether to introduce the innovation. Accordingly, the author sets up a framework of adoption decision highlighting roles of core managers, as showed in Fig. 4.1.

According to the process model of adoptive management innovation, the scheme selection is the direct result of the proposal evaluation, belonging to the natural process; the formation of innovation intention belongs to psychological activities of

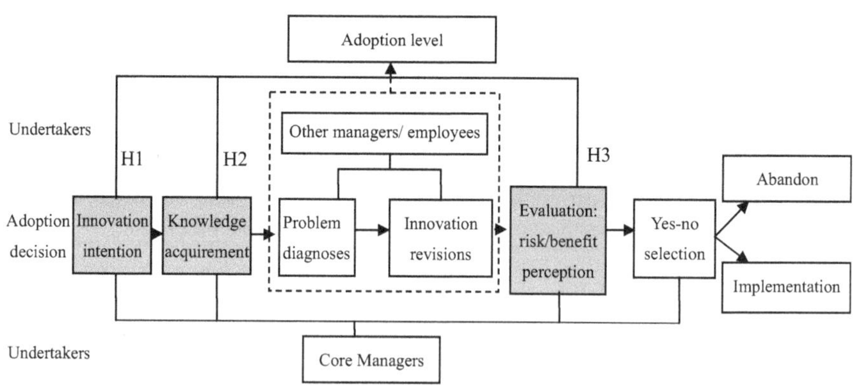

Fig. 4.1 The framework of adoption decision and the undertakers

the core managers; although the proposal evaluation theoretically requires top managers to objectively and scientifically make a comprehensive evaluation of the expected risks and benefits, characteristics (such as the ambiguity, complexity and the uncertainty of results) of management innovation make it more difficult to evaluate scientifically, and as a result more relying on managers' subjective perception, which also belongs to psychological activities. That is, the attitude formation and the proposal evaluation can only be accomplished by top managers individually, while the processes of problem identification, knowledge acquirement, problem diagnoses, innovation revisions and the implementation belong to more explicit practice activities. More specifically, the knowledge acquirement, problem diagnoses and innovation revisions can be called the innovation proposal creation. Even though the main responsibility body is still top managers of the organization, considering the limitations of knowledge and competence of top managers, this chapter proposes that enhancing the interaction between top and middle managers in problem identification and innovation proposal creation can effectively improve the effectiveness of innovation decision-making. In addition, given that the researches on two aspects of the implementation involve both respective roles, rather than the interactions. Therefore, it's necessary to further strengthen the interactions of them to improve the effectiveness of innovation implementation.

4.3 Three-Dimension Decision Model

According to the set-up framework, except for the last spontaneous stage which hardly refers to the characteristics of managers and the phase of diagnosing specific problems and establishing scheme undertaken by lower-level managers or employees, the rest three stages indicate that main activities of individual managers in the process of introducing new management practices refer to shaping innovation intention, acquiring knowledge and perceiving risks. In another word, managers with stronger innovation intention, higher levels of knowledge acquirement and lower levels of risk perception, tend to be more active in introducing new practices. Therefore, the author adopts the term of adoption level of new management practices to describe different preferences of managers in management innovation practices. When used for describing innovation state of an organization during a certain period in the empirical study, it can be measured by the number of new practices or methods adopted. Then, this research formally proposes that:

Hypothesis 1: The stronger innovation intention managers have, the higher adoption level of new management practices will be;

Hypothesis 2: The higher levels of knowledge acquirement managers have, the higher adoption level of new management practices will be;

Hypothesis 3: The higher levels of risk perception managers have, the lower adoption level of new management practices will be.

The established decision process framework indicates that innovation intention, knowledge acquirement and risk perception are major activities taken by core managers through the process of adoption decision. However, what elements are able to promote stronger intention, higher level of knowledge acquirement and lower risk perception is worthy of further exploration.

4.3.1 The Dimension of Entrepreneurial Orientation

Through an investigation into companies like Yili Group, Haier Group, China Merchants Bank, and visiting to some of their core managers, the author finds out a common characteristic the obvious timeliness, initiative and enthusiasm to cope with the pressure of external competition, the dilemma of the internal management and the entrance of new organizations in the enterprise management (Miller 1983; De Clercq et al. 2010), namely, entrepreneurial orientation, existing among these successful managers. An entrepreneurial individual is one that "engages in product market innovation, undertakes somewhat risky ventures, and is first to come up with 'proactive' innovations, beating competitors to the punch" (Miller 1983). That is, a manager with entrepreneurial orientation has strong propensity to adopt activities to struggle against stress from external environment, and remove internal operation obstacles (De Clercq et al. 2010). As for individual managers, entrepreneurial orientation indicates an attitude that may be able to strengthen innovation intention of individuals, so as to improve adoption level of new practices. This process is consistent with the theory of planned behavior proposed by Ajzen (1985, 1987) that activities of human beings go through a process from attitude to intention, and finally to performance of a given behavior.

As Miller (1983) argues, entrepreneurial orientation can be characterized and tested from dimensions of innovativeness, risk taking, and proactiveness. Numerous researchers have adopted an approach based on Miller's (1983) original conceptualization (e.g. Covin and Slevin 1989; Naman and Slevin 1993; Schafer 1990), so do us. The author suggests that all three dimensions may affect innovation intention toward new practices. Among them, innovativeness reflects a tendency to engage in and support new ideas, novelty, experimentation, and creative processes that may result in new products, services, or technological processes (Miller 1983; Lumpkin and Dess 1996), which only focuses on technological innovation, with few management innovation involved. Nonetheless, the synergy theory of technological and management innovation (Vickery et al. 1999) indicates that, if entrepreneurs in technology innovation show a stronger tendency, their wishes of management innovation will be more stronger. The reasons are from two aspects: processes and results of technological innovation not only stimulate the change of management patterns or methods, but also facilitate the adoption and implementation of new practices (Calantone and Stanko 2007). Moreover, the frequent technology innovation activities are conducive to cultivate creative thoughts of managers and employees, which make the management innovation better accepted by employees and implemented. It's also helpful for improving management innovation. Proactiveness refers to processes aimed at anticipating and acting on future needs by seeking new opportunities which may or may not be related to the present line of operations (Lumpkin and Dess 1996), introduction of new management methods or practices ahead of competition, strategically eliminating operations which are in the mature or declining stages of life cycle. The traditional A-U model proposed by Abernathy and Utterback (1978) shows that, when products and technologies turn out to be convergent, management innovation becomes the key for competition, and then the intention toward new practices becomes stronger. In fact, even when technologies are laggard, managers are declined to treat management innovation as an effective way to make up. Therefore, proactiveness can actively encourage entrepreneurs to produce a preference of new management innovation and take management innovation actions ahead of the competitions. Risk taking is used to describe the degree to which managers are willing to make large and risky resource commitments, i.e., those which have a reasonable chance of costly failures, or the willingness of individual managers to invest in projects, activities and decisions with high uncertainty (Lumpkin and Dess 1996). It reflects the trend of managers to get rid of the current situation and realize development in uncertainty (Lumpkin and Dess 1996; Moreno and Casillas 2008). Individuals with high risk propensity pay more attention to the potential benefit of risky behaviors, which would activate strong aspiration for taking risky behaviors. Though adoptive management innovation (the radical adjustment or change of the structure of the organization, management concept or the business process) may be implemented successfully somewhere in other areas, it is still highly risky due to the significant differences of specific situations among different organizations. Hence, individuals with high risk propensity are more likely to introduce management innovation actions. Thus, all three dimensions of entrepreneurial orientation may positively affect innovation intention of managers. Therefore, this research proposes that:

Hypothesis 4: Entrepreneurial orientation of managers is positively related to managers' innovation intention of introducing new practices.

4.3.2 The Dimension of Social Network

Besides innovation intention, the adoption of new management practices also refers to activities of knowledge acquirement and risk perception. Literatures show that, the social network theory enables to address the problem of how to acquire innovative knowledge, and the cognition theory is workable in uncovering the internal psychological process of risk perception (Shane and Venkataraman 2000). The social network theory points out that, both organizations and individuals are not isolated, but exist in complex networks. Similarly, members of the network are not always directly linked to each other, but are likely to exist multi-layer indirect relationships between each other. While managers own some innovation knowledge, they often complement their knowledge by accessing their contacts (Aldrich and Zimmer 1986; Cooper et al. 1995), namely through their social network which extends across professional networks, reaching friends, colleagues from earlier jobs, and so on. The construction of social network reflects a long-term social process (Jack 2005) through which members recognize common interests, acquire knowledge from each other and build up trust foundation, emphasizing on the social relations represented by the connection. Social network is a multidimensional concept (Lesser 2000; Nahapiet and Ghoshal 1998). According to the definition of Mitchell et al. (2004), the social network is a specific connection between members of a certain group, or a formal or informal connection between the core manager and other acquaintances; while in the view of organizational level, the social network is the connection relationship constituted by the core organization and its related organizations. Existing literatures point out that, its structural characteristics can be tested from perspectives of network size, network heterogeneity, link strength and structural holes.

Firstly, the total network size refers to the number of contacts that managers turn to when they search for knowledge about new practices and the amount of knowledge and information these contacts hold. The larger a manager's network is, the more contacts and knowledge he holds, and the broader and more diverse the available knowledge for him will be. That is, large network facilitates knowledge acquirement of managers by improving efficiency and cutting down time and cost (De Carolis and Saparito 2006). In the larger social network, entrepreneurs are more likely to contact with different types of knowledge, which makes entrepreneurs possible to filter a large number of obtained knowledge so that they can select innovation knowledge highly related to internal problems of the organization.

Secondly, obviously heterogeneous networks consist of members with different backgrounds, careers, ages and professionals (Burt 1992). Knowledge held by these different individuals is characterized by nonredundance and diversification, which indicates that heterogeneous networks enable to offer more knowledge sources to meet the demand of managers for diversified innovation knowledge. In addition, nonredundance may help to reduce time for acquiring knowledge from networks. Moreover, through the understanding between entrepreneurs and network members, when producing different knowledge demands, entrepreneurs can obtain reliable knowledge more accurately.

Thirdly, link strength of networks refers to the intimacy of net members as manifested in "strong" versus "weak" ties. The "strength" of a tie is a reflection of the combination of the amount of time, emotional intensity, intimacy, and reciprocal services that characterize that tie (Granovetter 1985). Weak ties represent connection relationships among groups owning different information sources, and always deliver a large amount of less repeated information and have a lower cost. Thereby, the more weak ties are, the wider tacit knowledge is, the more diverse the knowledge is, the newer knowledge entrepreneurs obtain, the higher the level of individual knowledge acquisition is. Though researchers like Granovetter (1973, 1982, 1985) argue that weak ties are more important in information searching because they provide a greater diversity of information, strong ties are typically associated with trust and facilitate the flow of fine-grained information (Gulati 1998; Rowley et al. 2015) and the transfer of tacit knowledge (Uzzi 1997). Information searchers are more interested in the quality rather than quantity of information, the diverse information obtained through weak ties may have no value. By contrast, in strong ties, people usually know more about each other so that they are easier to realize what information is more important to each other. So individuals with strong ties own advantage in knowing each other, recognizing important information for other members and offering helps (Granovetter 1982). That is, strong ties offer individuals with shortcuts for knowledge searching, cut down costs for supervision and negotiation, and raise the reliability of knowledge (Adler and Kwon 2002). That is to say, both weak ties and strong ties could promote management innovation knowledge acquisition.

Finally, a structural hole is defined as the separation between nonredundant contacts (Burt 1992). In fact, a structural hole is the lack of a tie between two actors. Another actor who bridges this hole, i.e., has relationships with the two actors who do not know one another, is theorized to have specific information benefits. Burt (1992) emphasize that, unrepeated ties really play important roles when information repeating problems exist between two individuals that know each other in the social network. Burt (1992) argues that networks filled with structural holes provide three information benefits: access, timing and referral, all of which are important for acquiring knowledge about new management practices. In a word, the more connected structural holes mean less contact or not contact between other members of the network and less redundancy or shared information, therefore, the knowledge an individual acquire has more diverse sources, stronger accessibility and timeliness as well as higher reliability.

Thus, all four dimensions of social network may positively affect knowledge acquirement of managers. Therefore, this research proposes that:

Hypothesis 5: Social network of managers is positively related to managers' knowledge acquirement in introducing new practices.

4.3.3 The Dimension of Cognitive Biases

Management innovation cognition involves the knowledge structure of managers in evaluating, making judgment and confirming new practices. As behavior decision research indicates, the cognition capability of individuals is bounded that they can neither acquire all relevant information nor understand information acquired completely. Bounded cognition accompanied with complex, vague and uncertain decision contexts, leads to the emergence of cognitive biases in human beings (Einhorn and Hogarth 1986). In the decision making process of adoptive management innovation, entrepreneur's cognitive biases is a phenomenon that entrepreneurs produce deviation to the cognitive of existing management innovation thoughts and practices and the actual situation under the influence of individual psychological or behavioral factors. The author does not intend to use the term "bias" as either a negative or positive state of affairs, but to use the term objectively to identify a manager's cognition and decision processes that lead to introduction of new practices. In brief, cognitive biases represent a simplified pattern adopted by individuals when confronted with uncertainty (Tversky and Kahneman 1974; Bazerman 1990; Busenitz and Lau 1996). Simon et al. (2000) suggest three specific cognitive biases that influence risk perception as they relate to managers: overconfidence, illusion of control, and representativeness.

Firstly, overconfidence refers to the failure to realize the limits of one's knowledge (Bazerman 1990; Lichtenstein and Fischhoff 1977; Oskamp 1965). In another word, overconfident individuals overestimate the probability of being right. Its psychological basis appears to be individuals' refusal to examine any unsubstantiated assumptions they might hold. Overconfidence managers tend to drive enterprises into new areas (Schwenk 1986) to engage in risk investment (McCarthy et al. 1993) or the important innovation practices. During the process of introducing a new practice, managers, for one aspect, tend to believe "I'm always right" which holds them back from admitting their mistakes of ignorance when make unsuitable evaluation or judgment on new practices; for another aspect, managers may ignore new information they receive after making an initial decision, or they do not realize the extent to which their estimates are inaccurate. Based on the two aspects, once managers make a positive response after mastering part of new management practice knowledge in the decision making process of management innovation, even if they obtain other negative information about the new management practice later, they still have to make their previous decisions and are not willing to adjust. Thus, overconfident managers may treat their assumptions as fact and believe that their introduction actions are less risky than they really are.

Secondly, illusion of control is defined as "an expectancy of a personal success probability inappropriately higher than the objective probability would warrant" (Langer 1975). Compared with overconfidence, overconfidence stresses on overestimating the accuracy of the personal knowledge or personal certainty to the current event, while illusion of control emphasizes to overestimate personal response and ability to predict the future events or results. This expectancy stems from the fact that people will seek out information that supports their opinions while ignoring contradictory information. In another word, illusion of control indicates that individuals tend to overestimate the skills of them, believe their control and prediction of the future uncertainty, and show a strong tendency of control to the results. By feeling that they can control and predict outcomes, individuals will overestimate the success possibility of their risky behavior (Duhaime and Schwenk 1985; Schwenk 1986), evaluate the hazards inherent in situations in a more favorable light (De Carolis and Saparito 2006), and subjectively improve the propensity of introducing new management practices. Consequently, it is visible that illusion of control lowers the level of risk perception in introducing new practices.

Thirdly, representativeness refers to a simplified thinking tactic that individually are willing to rely on small samples (March and Simon 1958) or a limited number of information sources to make a decision (Tversky and Kahneman 1974). Law of large Numbers claims that, only by large number of random samples can individuals reasonably come to the result in the statistical process. Although a small amount of samples can simplify the process of information processing to improve the efficiency, it's easy to get serious deviation of the information and facts (Kahneman and Lovallo 1993; Payne et al. 1992). However, since managers are always working in circumstances of information overload, high uncertainty, novel situations, strong emotions, time pressure, and fatigue (Baron 1998), they tend to use limited information to support their risky decisions (Baron 1998; Busenitz and Barney 1997; De Carolis and Saparito 2006). In addition, when individual mangers discuss new practices with a limited number of professors, advisors or colleagues, they are more likely to receive overly positive feedback (Kahneman and Lovallo 1993). In conclusion, acquiring management innovation knowledge from small samples, namely representativeness, will reduce the risk perception in introducing new practices.

Thus, all three cognitive biases may negatively affect risk perception of managers. Therefore, this research proposes that:

Hypothesis 6: Cognitive biases of managers are negatively related to managers' risk perception for introducing new practices.

Additionally, social network of managers may directly affect their cognitive biases and entrepreneurial orientation. As for the effect of social network on cognitive biases, for one aspect, large network size, high heterogeneity and a large number of structural holes produce psychological superiority for managers. Once the demand for resources and information emerges, managers have an intendancy to believe that their advanced social network would offer facilitation and even spiritual support. Meanwhile, in order to reduce uncertainty of innovation results, managers may treat social capital as a cushion. That is to say, individuals believe that they can continuously depend on their position advantage and information advantage in the social network to achieve the goal of control results and uncertainty (De Carolis and Saparito 2006). Making innovation and pursuing new opportunities at this time become more attractive. To efficiently acquire information and resources before others drives managers to overestimate their knowledge base in certain area and relevant techniques, which further produces overconfidence and illusion of control (De Carolis and Saparito 2006). For another aspect, strong ties may lead to cognitive biases. The trust produced by strong ties tends to create confident expectations about the future (Rousseau et al. 1998). Meanwhile, managers believe that information received from a trusted partner is more accurate and relevant (McEvily et al. 2003), so they may be less likely to verify the information's accuracy (McEvily et al. 2003) and may overestimate the information's probability of being right. The trust can also lead managers to only focus on the most close relationship members in the social network and the flow of their information, which limits wide flow and exchange of information. Coupled with not verifying the information's accuracy, managers consider only a small amount of information in the decision-making process, resulting in the appearance of representativeness. In the networks consisted of members with similar mental models, managers are easy to exaggerate the bear ability of management innovation, overestimate the influence of individual decision about the future, and lead to the illusion of control. Therefore, this research proposes that:

Hypothesis 7: Social network of managers is positively related to cognitive biases.

As for the effect of social network on entrepreneurial orientation, theory of interpersonal attraction indicates that individuals with similar faith tend to be attracted by each other, which would further strengthen shared attitude and behaviors. The responsibilities of managers decide that they have stronger tendency on innovativeness, risk propensity and proactiveness, that is, have a certain entrepreneurial orientation. Accordingly, social network of managers, which consists of a large number of interrelated members with common faith, value, personalities as well as the propensity to innovation or entrepreneurial orientation, may strengthen the effect of entrepreneurial orientation. Furthermore, social information processing theory also indicates that social effect plays an important role in attitude and behavior production of individuals. In the social network of managers, individuals with strong ties affect other members by exchanging and communicating and enable the network members to show significance in terms of entrepreneurial orientation and their effects passing in the whole network, which stimulates imitation and diffusion of characteristics like entrepreneurial orientation to further produce shared attitude and united behaviors, including entrepreneurial orientation, a positive attitude to cope with the enterprise competition. Therefore, this research proposes that:

Hypothesis 8: Social network of managers is positively related to entrepreneurial orientation.

In a word, the identified three major activities of managers including innovation intention, knowledge acquirement and risk perception are affected by entrepreneurial orientation, social network and cognitive biases, respectively. Accordingly, the author could deduce that direct effects of entrepreneurial orientation, social network and cognitive biases on adoption level of new management practices exist, as Hypothesis 9, 10 and 11. Then, the author builds up a three-dimension decision model of introducing new practices by integrating all direct and indirect affecting relationships and paths, shown as Fig. 4.2.

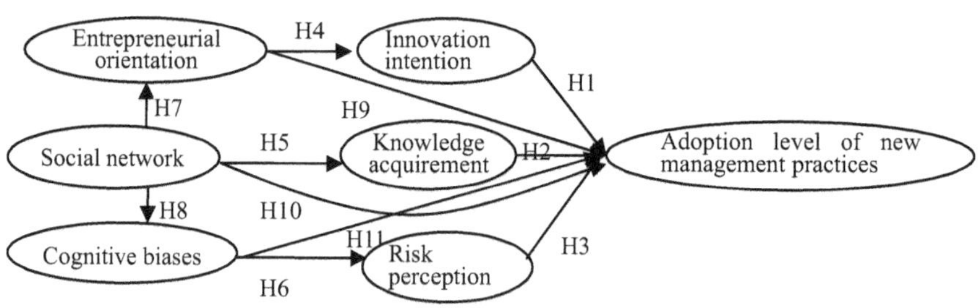

Fig. 4.2 The three-dimension decision model of adopting new practices

Hypothesis 9: Entrepreneurial orientation of managers is directly and positively related to adoption level of new practices.

Hypothesis 10: Social network of managers is directly and positively related to adoption level of new practices.

Hypothesis 11: Cognitive biases of managers are directly and negatively related to adoption level of new practices.

4.4 Method

4.4.1 Data Collection

The primary data collection vehicle was a survey emailed or mailed to core managers of firms in manufacturing industry whose tenure was over 3 years, in Zhejiang, Sichuan, Guangdong and Liaoning provinces, mainly through telephone reservation access, the alumni association and acquaintance to introduce. Here, core managers referred to CEO, owners of private enterprises, or other top level managers who have decision power for big issues as management innovation. The questionnaire for this survey was designed in an iterative manner according to the integrated research frameworks. The author started with an extensive literature review on management innovation and relevant research, then visited and interviewed 10 managers that had done well in introducing new management practices, and finally consulted with experts on these fields. Sending 724 questionnaires, the author totally recycled 306 questionnaires. In total, 237 of 724 managers from different firms concerned completed the survey, a response rate of over 32.7%, with 81.0% male respondents, 19.0% female; 56.1% aged between 31 and 40, 26.6% between 21 and 30, and 15.2% between 41 and 50; 61.2% managers having a company tenure between 3 and 5 years, 27.0% between 5 and 10 years, and 6.8% over 20 years; 48.5% of them holding master's degree or above, 27.0% with bachelor's degree and the less non well educated mainly referring to founders and directors of family businesses. In a word, respondents with these characteristics can represent general top level managers in China.

4.4.2 Measures

The questionnaire consisted of 7 parts: The first part investigated adoption level of management innovation, with 9 items from the nine aspects of investigation; The second part investigated entrepreneurial orientation by 9 items; The third part referred to innovation intention, including 5 items; The fourth part included 10 items referring to the social network to investigate the network size, structure holes, heterogeneity and link strength; The fifth part contained 7 items to investigate the level of knowledge acquisition; The sixth part consisted of 16 items concerning the cognitive biases; The seventh part set 5 risk perception items. The questionnaire has 61 questions in total (seen in appendix A).

All variables, except for adoption level of management innovation, network size, structural holes and overconfidence, were computed from the respondents' answers to five-point Likert scale-type items, with 1 = "strongly disagree" and 5 = "strongly agree".

(1) *Adoption level of management innovation*

According to the method adopted by Mol and Birkinshaw (2009) to measure introduction of new management practices and the classification of management innovation in The National Enterprises Management Modernization Innovation Achievement, the author measured this adoption level by asking respondents: "Did your firm make major changes in the following areas of business practices during the period 2007–2009? (1) Implementation of new or significantly changed corporate strategies; (2) Using more advanced management techniques; (3) Changing the organizational structure; (4) Implementation of business process reengineering; (5) Reforming the organizational culture; (6) Changing significantly your firm's marketing concepts/strategies; (7) Introducing new methods of risk-controlling and financial management; (8) Implementation of new human resource managements or channel management systems; (9) Other new practices (0 = 'not used'; 1 = 'used')". A single scale was applied with the value of "0" for no effective management innovation activity at all, with "1" added for each type of management innovation the firm engaged in, such that the maximum value is 9. The higher value a sample company held, a higher level of adopting new practices it had.

(2) *Entrepreneurial orientation*

Entrepreneurial orientation was measured as the mean scores of three dimensions each with three questionnaire items respectively based on the classic scales developed by Miller and Friesen (1982), Naman and Slevin (1993) and Wiklund (1999): (1) More attention has been paid to technological innovation; (2) A large number of technological innovation projects have been promoted during the last three years; (3) You prefer to radical changes; (4) You initiate actions to which competitors then respond; (5) You are always the first to introduce new products or management pattern; (6) You adopt a very competitive, undo-the-competitors posture; (7) There is a strong proclivity for high-risk projects; (8) Owning to the nature of the environment, bold, wide-ranging acts are necessary to achieve the firm's objectives; (9) You adopt a bold, aggressive posture in order to maximize the probability of exploiting potential opportunities. This variable's reliability was 0.859.

(3) *Innovation intention*

Without existing scales to directly measure innovation intention, innovation intention was measured as the mean score of five questionnaire items exploited by ourselves based on the definitions of management innovation intention and management innovation practice and the measure of entrepreneurial intention conducted by Bird (1988) and Krueger and Carsrud (1993): (1) You are considering introducing and implementing at least one new management practice now; (2) The existing operation or management system of your organization is temporal and changeable; (3) At least one practice will be introduced in the following year; (4) The idea of changing the existing operation or management system always exists; (5) The existing operation or management system needs to be improved dramatically. The reliability of this variable was 0.810.

(4) *Social network*

Social network was measured as the mean score of its four dimensions, including network size, structural holes, link strength and network heterogeneity.

Network size was measured by asking respondents to enter the initials of up to 10 people who are important sources of knowledge on new management practices. An upper limit 10 was chosen for two reasons: existing research suggests that managers' networks include fewer than 12 ties, for example, Carroll and Teo (1996) found that the mean network size among managers was 3.42 with a standard deviation of 1.70, and Seibert et al. (2001) found the number was 5.32 with a standard deviation of 1.99; asking mangers to give information on more than 10 people was judged to risk serious respondent fatigue. The number of structural holes was measured by adopting the method of calculating effective network size proposed by Tortoriello et al. (2004). Specifically, after listing the people they saw as important sources of information, they were asked to indicate whether or not these individuals know each other by circling the appropriate letter in the matrix. For example, if individuals one(1) and two(2) know each other then circle "Yes" in the cell located at row 1 and column 2. Average link strength was operationalized according to the three elements that Krackhardt (1992) discussed as being critical: interaction, affection and history of interaction. For each link managers listed, they needed to answer three items to measure link strength. The average link strength for each manager was then created by taking the average of the link strength for each link. Network heterogeneity was measured as a mean score of four questionnaire items designed by myself based on its definition.

(5) *Knowledge acquirement*

Knowledge acquirement was measured as a mean score of seven questionnaire items based on Huber (1991) and Soo et al. (2004): (1) new methods for internal operation could be obtained as soon as possible; (2) you have various channels for new management practices; (3) knowledge from different channels is heterogeneous; (4) the available innovation knowledge could satisfy the need for handling management problems; (5) the obtained innovation knowledge is always reliable; (6) you are able to acquire the latest knowledge on new practices; (7) you can always get knowledge you need. The reliability of this variable was 0.846.

(6) *Cognitive biases*

The value of cognitive biases was measured as the mean score of its three dimensions, including overconfidence, illusion of control and representativeness.

In order to measure overconfidence, managers were asked to answer 10 questions with A and B selections: (1) Which year was Lean Production born? (2) Who was the first one proposing the concept of Learning Organization? (3) Where was Total Quality Management born? (4) How many cars were produced in China in 2009? (5) What percentage of enterprises is middle and small in China in 2009? (6) What is the position of Gross National Product of China in 2009 around the world? (7) What is the growth rate of Chinese economy in 2009? (8) What is the average distance between the Moon and the Earth? (9) How long is the Yellow River? (10) What is the total population of Shanghai by the end of 2009? And then managers would select confidence level after answering each question, with "50% = complete uncertainty" to "100% = complete certainty". The confidence value of individuals was measured as the balance between average confidence level and right answer percent, with "0" as a demarcation point.

Illusion of control was measured as the mean score of five questionnaire items based on Thompson et al. (1998) and its characteristics of propensity to control the results and over the average. The five items were designed from perspectives of five problems managers concerned most referring to success rate, cost, acceptance rate of employees, timing and predicted effect of introducing new practices. The five items included: (1) Suppose the average success rate of adopting a new management practice is 50%, you believe that you are able to make it over 50%; (2) Suppose the average cost of adopting a new management practice is 1 million, you believe that you are able to make it less than 1 million; (3) Suppose that average 50% of employees would accept a new management practice, you believe that you are able to make it more than 50%; (4) Suppose that average time for implementing a new management practice is 3 years, you believe that you are able to make it less than 3 years; (5) Suppose that average success rate of a new management practice is 50%, you believe that you are able to make it more than 50%.

Combined with the practice of the management innovation, representativeness was measured through a single-item question relevant to specific situation that whether or not managers would decide to introduce the high-failure-rate practice of business process reengineering when two closed companies had successfully implemented it.

(7) *Risk perception*

Risk perception was measured as a mean score of five questionnaire items based on a contextual question (e.g. Simon et al. 2000). Respondents were asked to answer: "Suppose you are in charge of a large telecom operator, facing the challenges of both fierce competition press from the external environment and low efficiency of existing processes, you believe that: (1) the overall risk of introducing concept of Business Process Reengineering (BPR) to radically change the existing processes is high; (2) the probability of failure is very high; (3) the amount your company could lose is substantial; (4) introducing BPR will have a negative ramification for your company; (5) there is a high probability of your company losing a great deal by introducing BPR. The reliability of this variable was 0.865.

4.4.3 Model Approach

Structural Equation Modeling (SEM), a major component of applied multivariate statistical analyses, was adopted. Once a theory had been proposed, it could then be tested of theoretical model by SEM. Namely, SEM is a statistical technique for testing theoretical models or proposed hypotheses, which referred to as the confirmatory aspect. Accordingly, it's an appropriate method to demonstrate the causal relationships among entrepreneurial orientation, innovation intention, social network, knowledge acquirement, cognition biases, risk perception and adoption level of new practices, as well as the proposed models in this research. Besides, another aspect of SEM is the so-called exploratory mode, which allows for theory development and model improvement. Therefore, by using SEM, the established decision model could be tested and improved.

4.5 Analysis

4.5.1 Model Modification

The author used Amos18.0 to test the decision model. Table 4.2 presented the results of structural equation modeling analysis, which indicated that the built

Table 4.2 The fitness index of models

	The original model	The modified model
CMIN	334.706	194.756
DF	82	77
P	0.000	0.000
CMIN/DF	4.082	2.529
NFI	0.794	0.896
GFI	0.847	0.912
TLI	0.757	0.901
CFI	0.832	0.894
RMR	0.119	0.052
RMSEA	0.105	0.048
AIC	410.706	260.756
CAIC	580.493	452.883

decision model did not fit the data adequately ($\chi^2(82) = 334.706$, $\chi^2/df = 4.082 > 2$, $P < 0.001$, RMR = 0.119, RMSEA = 0.105, NFI = 0.794, GFI = 0.847, TLI = 0.757, CFI = 0.832, AIC = 410.706, CAIC = 580.493). Therefore, it was necessary to modify the original model.

The modified index (MI) of the original model indicated that affecting paths might exist from cognitive biases to entrepreneurial orientation (MI = 41.518), from risk perception to innovation intention (MI = 24.185), from knowledge acquirement to innovation intention (MI = 21.145) as well as from entrepreneurial orientation to knowledge acquirement (MI = 13.511). Therefore, the author added one path each time and tested the new model accordingly, with the decision model improved step by step. Table 4.2 presented the fitness index of the finally modified model, which indicated that the modified model fit the data adequately ($\chi^2(77) = 194.756$, $\chi^2/df = 2.529 > 2$, $P < 0.001$, RMR = 0.052, RMSEA = 0.048, NFI = 0.896, GFI = 0.912, TLI = 0.891, CFI = 0.874, AIC = 260.756, CAIC = 452.883).

4.5.2 *Coefficients Estimation*

Table 4.3 presented the path coefficients outputs to the modified decision model of adopting management innovation, and Fig. 4.3 presented the affecting paths of the modified model. The results confirmed all proposed relationships.

4.5.3 *Effect Analysis*

In order to further specify the relationships among variables, the author used AMOS18.0 to calculate the direct, indirect and total effects of all variables on introduction of new practices, presented in Table 4.4.

4.6 Discussion

The results of analysis confirmed the direct effects of innovation intention, knowledge acquirement, and risk perception on introduction of new management practices, with coefficients of 0.359, 0.543 and −0.646 respectively, supporting Hypotheses 1, 2 and 3. When managers have stronger intention to innovation, higher level of knowledge acquirement or perceiving lower risk, they may have a stronger intendancy to introduce new management practices. In another word, the results proved the reliability of decision process model proposed in the second part of this book based on both relevant literature and innovation practices.

Table 4.3 The path coefficients output to the modified decision model

			Coefficient	S.E.	C.R.	P	Standard coefficient
Cognitive biases	<—	Social network	0.321	0.077	4.194	***	0.509
Entrepreneurial orientation	<—	Social network	0.206	0.051	4.042	***	0.224
Entrepreneurial orientation	<—	Cognitive biases	0.384	0.058	6.626	***	0.362
Knowledge acquirement	<—	Social network	0.528	0.049	6.672	***	0.532
Risk perception	<—	Cognitive biases	−0.919	0.197	−4.674	***	−0.490
Knowledge acquirement	<—	Entrepreneurial orientation	0.377	0.084	4.472	***	0.326
Innovation intention	<—	Entrepreneurial orientation	1.447	0.230	6.299	***	0.692
Innovation intention	<—	Risk perception	−0.220	0.057	3.864	***	−0.245
Innovation intention	<—	Knowledge acquirement	0.225	0.054	4.174	***	0.238
Adoption level	<—	Innovation intention	0.665	0.169	3.944	***	0.359
Adoption level	<—	Knowledge acquirement	1.674	0.413	4.056	***	0.543
Adoption level	<—	Risk perception	−0.810	0.207	−3.915	***	−0.646
Adoption level	<—	Cognitive biases	1.668	0.344	4.836	***	0.158
Adoption level	<—	Social network	0.211	0.038	5.553	***	0.124
Adoption level	<—	Entrepreneurial orientation	1.791	0.404	4.428	***	0.207

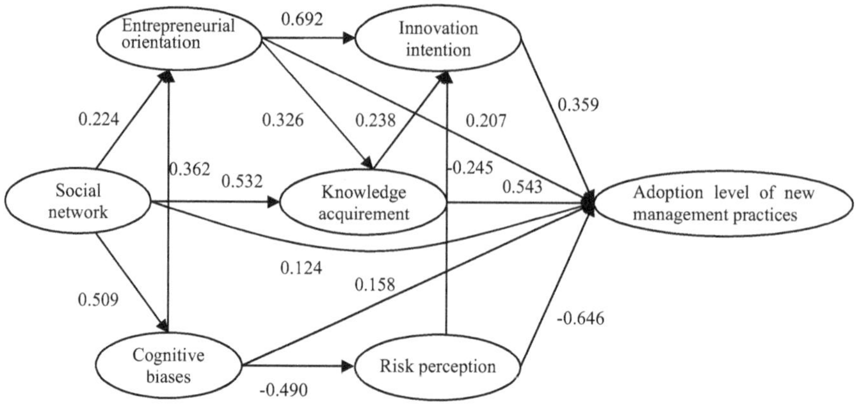

Fig. 4.3 Affecting paths of introduction decision model

The proposed positive effects of entrepreneurial orientation and social network on innovation intention and knowledge acquirement respectively (Hypotheses 4 and 5), and the negative effect of cognitive biases of managers on risk perception (Hypothesis 6) were supported (b = 0.692, 0.532, −0.490, $p < 0.001$). That is, besides the direct effects, entrepreneurial orientation, social network and cognitive biases of managers also indirectly affected introduction of new management practices through innovation intention, knowledge acquirement and risk perception, respectively. According to existed researches, when the indirect effect is larger than the direct effect, the influence of intermediary variables play an important role, and the intermediary role can't be ignored at this time. The indirect effect of entrepreneurial orientation on adoption level through innovation intention was 0.248 (0.692 * 0.359), and its direct effect was 0.207 (0.248 > 0.207), which indicated that innovation intention played an important mediate role between the two. Similarly, the indirect effect of social network on adoption level through knowledge acquirement was 0.289 (0.532 * 0.543), and its direct effect was 0.124 (0.289 > 0.124), which indicated that knowledge acquirement played an important mediate role between the two; the indirect effect of cognitive biases on adoption level through risk perception was 0.317 (−0.490 * −0.646), and its direct effect was 0.158 (0.289 > 0.124), which indicated that risk perception played an important mediate role between the two.

Besides the three relatively independent paths to adoption of new management practices, interrelations also exist among entrepreneurial orientation, social network and cognitive biases of managers as well as their three main activities in decision process including innovation intention shaping, knowledge acquirement and risk perception: (1) The proposed positive effects of social network on entrepreneurial orientation and cognitive biases (Hypotheses 7 and 8) were supported (b = 0.224, −0.509, $p < 0.001$). Accordingly, the author could deduce that cognitive biases mediated the effect of social network on entrepreneurial orientation. This finding offered an valuable insight that, it would be easier for managers to change or adjust

Table 4.4 The standard effect output of the model

		Social network	Cognitive biases	Entrepreneurial orientation	Risk perception	Knowledge acquirement	Innovation intention
Cognitive biases	Direct effect	0.509	0.000	0.000	0.000	0.000	0.000
	Indirect effect	0.000	0.000	0.000	0.000	0.000	0.000
	Total effect	0.509	0.000	0.000	0.000	0.000	0.000
Entrepreneurial orientation	Direct effect	0.224	0.362	0.000	0.000	0.000	0.000
	Indirect effect	0.184	0.000	0.000	0.000	0.000	0.000
	Total effect	0.408	0.362	0.000	0.000	0.000	0.000
Risk perception	Direct effect	0.000	−0.490	0.000	0.000	0.000	0.000
	Indirect effect	−0.250	0.000	0.000	0.000	0.000	0.000
	Total effect	−0.250	−0.490	0.000	0.000	0.000	0.000
Knowledge acquirement	Direct effect	0.532	0.000	0.326	0.000	0.000	0.000
	Indirect effect	0.133	0.118	0.000	0.000	0.000	0.000
	Total effect	0.665	0.118	0.326	0.000	0.000	0.000
Innovation intention	Direct effect	0.000	0.000	0.692	−0.245	0.238	0.000
	Indirect effect	0.501	0.399	0.076	0.000	0.000	0.000
	Total effect	0.501	0.399	0.770	−0.245	0.238	0.000
Adoption level	Direct effect	0.124	0.158	0.207	−0.646	0.543	0.359
	Indirect effect	0.865	0.599	0.453	−0.088	0.085	0.000
	Total effect	0.989	0.757	0.660	−0.734	0.628	0.359

structural characteristics of their implicit social network in the process of improving adoption level of new practices than change entrepreneurial orientation and cognitive biases which were explicit. Thus, managers could achieve the goal of changing cognitive biases and entrepreneurial orientation and ultimately affecting the management innovation decision by changing the social network; (2) The modified positive effect of cognitive biases on entrepreneurial orientation was supported (b = 0.362, $p < 0.001$), which showed that cognitive biases of overconfidence, illusion of control and representativeness might be major causes for technology innovativeness, proactiveness and risk propensity of managers. In another word, considering that a number of studies had verified the entrepreneurial orientation could actively improve enterprise performance, this confirmed that cognitive biases were not always negative, but might play a positive role in decision process. Mangers with certain degree of cognitive biases had an intendancy to break routines and hidebound thoughts as well as pursue new practices and methods, and finally achieve the new goal of the organization; (3) The modified positive effect of entrepreneurial orientation on knowledge acquirement was supported (b = 0.326, $p < 0.001$), which suggested that, as a conscious activity, knowledge acquiring would be affected not only by implicit factors like channels and means but also by initiative of managers. That is to say, in the process of taking management innovation practices ahead of competitors, in order to meet the demand for knowledge, managers would take the initiative to search and acquire knowledge; (4) The results also showed that, innovation intention would be affected by the risk perceived and knowledge acquired. The higher the risk perception was, the lower the innovation intention would be, and the higher the level of knowledge acquisition was, the stronger the innovation intention would be. This was consistent with the argument proposed by Sternberg and Lubart (1991) that innovation intention was affected by multiple elements including intelligence, knowledge and cognition. In addition, entrepreneurial orientation would help managers improve their levels of knowledge acquisition, owing to the fact that entrepreneurial orientation provided motivations for managers to acquire knowledge.

Effects of three major dimensions on adoption of new practices include both direct and indirect paths. The direct effects of entrepreneurial orientation, social network and cognitive biases on adoption level were 0.207, 0.124 and 0.158 respectively, which supported Hypotheses 9, 10 and 11, and indirect effects were 0.453, 0.865 and 0.599, even exceeding direct effects. Owing to interrelations existing among variables besides three independent paths, more indirect effects emerged in this decision model, for example, knowledge acquirement partly mediated the effect of entrepreneurial orientation on adoption level, cognitive biases might directly affect entrepreneurial orientation and indirectly affect innovation intention, knowledge acquirement and even adoption of new practices. Therefore, adoption decision of new practices was a complex process. The total effects of entrepreneurial orientation, social network and cognitive biases on introduction of new practices were 0.660, 0.989 and 0.757, which indicated that, with other variables constant, when entrepreneurial orientation, social network and cognitive biases increased by one percent, adoption of new practices would increase by 0.660,

0.989 and 0.757 percent respectively. Additionally, when all three dimensions had been proved to be major factors of introducing new practices, the role of social network was most important.

Understanding why some managers pursue innovative opportunities by introducing new management practices while others do not is a major focus in the study of adoptive management innovation in China. This research contributes to this field by proposing that entrepreneurial orientation, social network and cognition of managers are important in understanding the adoption of new practices. By showing specific affecting paths of three major characteristics of managers, three main activities abstracted from the practices and literatures of management innovation, this section aims to unveil the black box in complex decision process.

4.7 Conclusion

The decision model holds important implications for management practices aiming to improve management efficiency by introducing new ideas or methods. Again, the critical roles that entrepreneurial orientation, social networks and cognitive biases play in the introduction of new practices as well as their affecting paths need to be reexamined. Although the author does not address the ultimate success or failure of a new practice, however that might be measured, surely unrealistic introduction of new practices result in wasted time, energy, and resources.

However, the author explores the decision mechanism of adopting new practices from a single perspective, namely the innovation decision makers of firms. The research does not investigate other variables that may affect innovation decision, such as contextual elements like competitive intensity, internal elements like structure and culture, as well as top management teams. Future research may integrate more variables to deepen the study on the internal mechanism of introducing new practices. Particularly, in the context of China, most enterprises tend to introduce existing management concepts and methods from the external, but not to create new practices by themselves, when confronted with the disadvantages of short development history, weak financial base, and fierce competitive press. It may be more significant to address the problem of how managers make essential decisions like adoption of new practices here in China.

References

Abernathy, W.J., Utterback, J.M.: Patterns of industrial innovation. Technol. Rev. **80**(7), 40–47 (1978)

Adler, P.S.: Social capital: prospects for a new concept. Acad. Manag. Rev. **27**(1), 17–40 (2002)

Ajzen, I.: From Intentions to Actions: A Theory of Planned Behavior. Springer, Berlin Heidelberg (1985)

Ajzen, I.: Attitudes, traits, and actions: dispositional prediction of behavior in personality and social psychology. In: Berkowitz, L. (ed.) Advances in Experimental Social Psychology, pp. 1–63. Academic Press 20, New York (1987)

Aldrich, H.E., Zimmer, C.: Entrepreneurship through social networks. (1986)

Barley, S.R., Kunda, G.: Design and devotion: surges of rational and normative ideologies of control in managerial discourse. Adm. Sci. Q. **37**(3), 363–399 (1992)

Baron, R.A.: Cognitive mechanisms in entrepreneurship: why and when enterpreneurs think differently than other people. J. Bus. Ventur. **13**(4), 275–294 (1998)

Bazerman, M.: Judgment in Managerial Decision Making. Wiley, New York (1990)

Bird, B.: Implementing entrepreneurial ideas: the case for intention. Acad. Manag. Rev. **13**(3), 442–453 (1988)

Burt, R.S.: Structural Holes: The Social Structure of Competition. Harvard University Press, Cambridge, MA (1992)

Busenitz, L.W., Barney, J.B.: Differences between entrepreneurs and managers in large organizations: biases and heuristics in strategic decision-making. J. Bus. Ventur. **12**(1), 9–30 (1997)

Busenitz, L.W., Lau, C.M.: A cross-cultural cognitive model of new venture creation. Entrep. Theory Pract. **20**(4), 25–39 (1996)

Calantone, R.J., Stanko, M.A.: Drivers of outsourced innovation: an exploratory study. J. Prod. Innov. Manag. **24**(3), 230–241 (2007)

Carroll, G.R., Teo, A.C.: On the social networks of managers. Acad. Manag. J. **39**(2), 421–440 (1996)

Cooper, A.C., Folta, T.B., Woo, C.Y.: Entrepreneurial information search. J. Bus. Ventur. **10**(2), 107–120 (1995)

Covin, J.G., Slevin, D.P.: Strategic management of small firms in hostile and benign environments. Strateg. Manag. J. **10**(1), 75–87 (1989)

Cyert, R.M., March, J.G.: A Behavioral Theory of The Firm. Prentice-Hall, Englewood Cliffs, NJ (1963)

De Carolis, D.M., Saparito, P.: Social capital, cognition, and entrepreneurial opportunities: a theoretical framework. Entrepreneurship Theory Pract. **30**(1), 41–56 (2006)

De Clercq, D., Dimov, D., Thongpapanl, N.: The moderating impact of internal social exchange processes on the entrepreneurial orientation-performance relationship. J. Bus. Ventur. **25**(1), 87–103 (2010)

Duhaime, I.M., Schwenk, C.R.: Conjectures on cognitive simplification in acquisition and divestment decision making. Acad. Manag. Rev. **10**(2), 287–295 (1985)

Einhorn, H.J., Hogarth, R.M.: Judging probable cause. Psychol. Bull. **99**(1), 3–19 (1986)

Elenkov, D.S., Judge, W., Wright, P.: Strategic leadership and executive innovation influence: an international multi-cluster comparative study. Strateg. Manag. J. **26**(7), 665–682 (2005)

Granovetter, M.: The strength of weak ties: a network theory revisited. In: Marsden, P.V., Lin, N. (eds.) Social Structure and Network Analysis, pp. 105–130. Sage Publications, Beverley Hills (1982)

Granovetter, M.: Economic action and social structure: a theory of embeddedness. Am. J. Sociol. **91**(3), 481–510 (1985)

Granovetter, M.S.: The strength of weak ties. Am. J. Sociol. **78**(6), 1360–1380 (1973)

Guillén, M.F.: Models of Management: Work, Authority, and Organization in A Comparative Perspective. University of Chicago Press, Chicago (1994)

Gulati, R.: Alliances and networks. Strateg. Manag. J. **19**(4), 293–317 (1998)

Hamel, G.: The why, what, and how of management innovation. Harvard Bus. Rev. **84**(2), 72–84 (2006)

Hashem, G., Tann, J.: The adoption of ISO 9000 standards within The Egyptian Context: a diffusion of innovation approach. Total Qual. Manag. Bus. Excellence **18**(6), 631–652 (2007)

Huber, G.P.: Organizational learning: the contributing processes and the literatures. Organ. Sci. **2**(1), 88–115 (1991)

Jack, S.L.: The role, use and activation of strong and weak network ties: a qualitative analysis. J. Manage. Stud. **42**(6), 1233–1259 (2005)
Kahneman, D., Dan, L.: Timid choices and bold forecasts: a cognitive perspective on risk taking. Manage. Sci. **39**(1), 17–31 (1993)
Kimberly, J.R., Evanisko, M.J.: Organizational innovation: the influence of individual, organizational, and contextual factors on hospital adoption of technological and administrative innovations. Acad. Manag. J. Acad. Manag. **24**(4), 689–713 (1981)
Krackhardt, D.: The strength of strong ties: the importance of philos in organizations. In: Nohria, N., Eccles, R.G. (eds.) Networks and Organization. Harvard Business School Press, Boston (1992)
Krueger, N.F., Carsrud, A.L.: Entrepreneurial intentions: applying the theory of planned behaviour. Entrepreneurship Reg. Dev. **5**(4), 315–330 (1993)
Langer, E.J.: The illusion of control. J. Pers. Soc. Psychol. **32**(2), 311–328 (1975)
Lesser, E.L.: Leveraging social capital in organizations. In: Lesser, E.L. (ed.) Knowledge and Social Capital: Foundations and Applications, pp. 3–16. Butterworth-Heineman, Boston (2000)
Lichtenstein, S., Fischhoff, B.: Do those who know more also know more about how much they know? Organ. Behav. Hum. Perform. **20**(2), 159–183 (1977)
Lincoln, Y.S., Guba, E.G.: Naturalistic Inquiry Sage. Naturalistic Inquiry. Sage Publications, California (1985)
Lumpkin, G.T., Dess, G.G.: Clarifying the entrepreneurial orientation construct and linking ii to performance. Acad. Manag. Rev. **21**(1), 135–172 (1996)
March, J.G., Simon, H.A.: Organizations. Wiley, New York (1958)
Mccarthy, A.M., Schoorman, F.D., Cooper, A.C.: Reinvestment decisions by entrepreneurs: rational decision-making or escalation of commitment? J. Bus. Ventur. **8**(1), 9–24 (1993)
Mcevily, B., Perrone, V., Zaheer, A.: Trust as an organizing principle. Organ. Sci. **14**(1), 91–103 (2003)
Miller, D.: The correlates of entrepreneurship in three types of firms. Manage. Sci. **29**(7), 770–791 (1983)
Miller, D., Friesen, P.H.: Innovation in conservative and entrepreneurial firms: two models of strategic momentum. Strateg. Manag. J. **3**(1), 1–25 (1982)
Mitchell, R.K., Busenitz, L., Lant, T., Mcdougall, P.P., Morse, E.A., Smith, J.B.: The distinctive and inclusive domain of entrepreneurial cognition research. Entrepreneurship Theory Practi. **28** (6), 505–518 (2004)
Mol, M.J., Birkinshaw, J.: The sources of management innovation: when firms introduce new management practices. J. Bus. Res. **62**(12), 1269–1280 (2009)
Moreno, A.M., Casillas, J.C.: Entrepreneurial orientation and growth of SMEs: a causal model. Entrepreneurship Theory Practi. **32**(3), 507–528 (2008)
Nahapiet, J., Ghoshal, S.: Social capital, intellectual capital, and the organizational advantage. Acad. Manag. Rev. **23**(2), 242–266 (1998)
Naman, J.L., Slevin, D.P.: Entrepreneurship and the concept of fit: a model and empirical tests. Strateg. Manag. J. **14**(2), 137–153 (1993)
Ocasio, W.: Towards an attention-based theory of the firm. Strateg. Manag. J. **18**(S1), 187–206 (1997)
Oskamp, S.: Overconfidence in case-study judgments. J. Consult. Psychol. **29**(3), 261–265 (1965)
Payne, W.A., Malcolm, D.C., Hossner, L.R., Lascao, R.J., Onken, A.B., Wendt, C.W.: Soil phosphorus availability and pearl millet water-use efficiency. Cropence **32**(4), 1010–1015 (1992)
Rousseau, D.M., Sitkin, S.B., Burt, R.S., Camerer, C.: Introduction to special topic forum: not so different after all: a cross-discipline view of trust. Acad. Manag. Rev. **23**(3), 393–404 (1998)
Rowley, T., Behrens, D., Krackhardt, D.: Redundant governance structures: an analysis of structural and relational embeddedness in the steel and semiconductor industries. Strateg. Manag. J. **21**(3), 369–386 (2015)

Schafer, D.S.: Level of entrepreneurship and scanning source usage in very small businesses. Entrepreneurship Theory Practi. **15**(2), 19–31 (1990)
Schwenk, C.R.: Information, cognitive biases, and commitment to a course of action. Acad. Manag. Rev. **11**(2), 298 (1986)
Seibert, S.E., Kraimer, M.L., Liden, R.C.: A social capital theory of career success. Acad. Manag. J. **44**(2), 219–237 (2001)
Shane, S., Venkataraman, S.: The promise of entrepreneurship as a field of research. Acad. Manag. Rev. **25**(1), 217–226 (2000)
Simon, M., Houghton, S.M., Aquino, K.: Cognitive biases, risk perception, and venture formation: how individuals decide to start companies. J. Bus. Ventur. **15**(2), 113–134 (2000)
Soo, Y.O., Cheng, Y., Wong, R., Hui, D.S., Lee, C.K., Tsang, K.K., Ng, M.H., Chan, P., Cheng, G., Sung, J.J.: Retrospective comparison of convalescent plasma with continuing cigh-dose methylprednisolone treatment in SARS patients. Clin. Microbiol. Infect. **10**(7), 676–678 (2004)
Sternberg, R.J., Lubart, T.I.: An investment theory of creativity and its development. Hum. Dev. **34**(1), 1–31 (1991)
Thompson, S.C., Armstrong, W., Thomas, C.: Illusions of control, underestimations, and accuracy: a control heuristic explanation. Psychol. Bull. **123**(2), 143 (1998)
Tortoriello, M., Mcevily, B., Perrone, V.: The evolution of status hierarchies: network dynamics and status differentiation. GSIA Working Paper (2004)
Tversky, A., Kahneman, D.: Judgment under uncertainty: judgment under uncertainty: heuristics and biases. Science **185**(4157), 1124–1130 (1974)
Uzzi, B.: Social structure and competition in interfirm networks: the paradox of embeddedness. Adm. Sci. Q. **42**(1), 35–67 (1997)
Vickery, S., Dröge, C., Germain, R.: The relationship between product customization and organizational structure. J. Oper. Manag. **17**(4), 377–391 (1999)
Wiklund, J.: The sustainability of the entrepreneurial orientation-performance relationship. Entrepreneurship Theory Practi. **24**(1), 37–48 (1999)
Williams, L., Rao, K.: Information technology adoption: using classical adoption models to predict AEI software implementation. J. Bus. Logistics (1998)
Yin, R.K.: Case study research: design and methods, 2nd edn. Sage, Newbury Park, CA (1994)
Zaltman, G., Duncun, R., Holbek, J.: Innovations and Organizations. Wiley, New York (1973)

Chapter 5
How Should Firms Deeply Implement Adoptive Management Innovation?

5.1 Introduction

During the past few decades, more and more firms around the world have actively involved in innovation, especially adopting new management practices (management innovation), aiming to realize their critical contributions to the success of firms through the creation and implementation of a new operating model, a management concept and method, an organizational structure, or a resource allocation mode (e.g. Birkinshaw and Mol 2006; Birkinshaw et al. 2007, 2008; Hamel 2006, 2007; Lin and Su 2010, 2014; Mol and Birkinshaw 2009; Vaccaro et al. 2012). Such management innovation reflects intentional reversals of personnel and structural growth dynamics in organizations and assists in producing, improving quality and enduring competitive advantage for continuous development (Pil and MacDuffie 1996). However, in most cases, management innovation fails to meet the intended objectives (Marks 2006; Paper and Chang 2005; Quinn 2004). A recent survey reports that only one-third of organizational innovation initiatives are considered successful by their organizational executives (Meaney and Pung 2008). Therefore, it has become an urgent task to improve performance of management innovation.

In recent years, researchers have paid much attention to what antecedents may affect management innovation. Antecedents, such as institutional factors (Guillén 1994; DiMaggio and Powell 1983), leadership behaviors (Vaccaro et al. 2012), employees (e.g. Kotter and Cohen 2002; Van et al. 2006; Whalen-Berry et al. 2003), the interaction between context and search (Mol and Birkinshaw 2009), and dynamic capabilities (e.g. Lin et al. 2016; Barney 1991; Teece et al. 1997; Wernerfelt 1984), have been considered or confirmed as drivers of management innovation. Most of these studies focus on effects of explicit elements, rather than implicit elements, while recessive traits (for example, social network, entrepreneurial orientation and cognitive deviation, etc.) may produce a more significant effect (Lin and Su 2010). Also, researchers have investigated into the process through which an innovation occurs to seek for the roots of their failures (e.g.

H. Lin, *Adoptive Management Innovation*,
https://doi.org/10.1007/978-981-10-7712-8_5

Birkinshaw et al. 2008; Lin and Su 2014). Different process frameworks have been identified, such as, evaluation, initiation, implementation and routinization (Hage 1980); initiation, adoption (decision) and implementation (Rogers 1995; Pierce and Delbecq 1977; Damanpour and Schneider 2006); motivation, invention, implementation, and theorization and labeling (Birkinshaw et al. 2008); or adoption decision and implementation (Lin and Su 2014). Though implementation has been identified as an indispensable phase without which firms are unable to realize the value of innovations (e.g. Birkinshaw et al. 2008; Lin and Su 2014), most research considers it as a natural action which does not need much attention. Similarly, though Chap. 3 of this book stresses the importance of innovation implementation and proposes a three-indispensably-interrelated-phase process for implementation, referring to preparation, overall implementation and solidification, it's a generating process extracted from explicit activities taken by managers or firms. In fact, management innovation is always triggered by a change of environment and driven by these antecedents; its intention is to recreate routines of an organization (McCabe 2002). Or, management innovation potentially requires fundamental changes in the routines of an organization, which reflects the difficulty and complication of innovation implementation (Argyris and Schän 1978; Birkinshaw et al. 2008).

Organizational routines, defined as repetitive, recognizable patterns of interdependent actions involving multiple actors (Feldman and Pentland 2003, p. 96; Howard-Grenville 2005, p. 618; Parmigiani and Howard-Grenville 2011, p. 417), have been commonly regarded as a central element of organizations and a fundamental mechanism for coordinating work therein (Feldman and Pentland 2003; March and Simon 1958; Cyert and March 1963; Cohen and Klepper 1996; Ortmann 2010). Traditional research treat routines as a source of stability in organizations which are designed to provide generalized solutions to recurring problems (Cyert and March 1963; March and Simon 1958; Ashforth and Fried 1988); while more recent research find that organizational routines actually instill flexibility in the face of changing external conditions (Feldman 2000; Feldman and Pentland 2003; Howard-Grenville 2005; Pentland and Rueter 1994; Rerup and Feldman 2011; Turner and Rindova 2012), which makes organizational routines a basic unit of analysis for organizational innovation and even a source of innovation (Feldman and Pentland 2003). This transformation of the debate on routines results from the rise of the practice-based perspective on organizational routines (Parmigiani and Howard-Grenville 2011). Accordingly, the implementation of management innovation refers to a project of overturning existing organizational routines and recreating new ones, which deeply involves the most micro level of an organization. To be noted, no matter generating management innovation or adoptive management innovation, their implementation represents a complicate process of recreating organizational routines. Therefore, in this Chapter, management innovation refers to both.

This chapter focuses on implementation mechanism of management innovation through an in-depth investigation into the micro-level of organizational routines, thereby addressing both theoretical and practical gaps with two questions. First, how firms implement management innovation through recreation of organizational routines? Management innovation is not a simple process of changing existing structures, processes or methods, but a complex one of fundamentally changing routines (Argyris and Schän 1978; Birkinshaw et al. 2008). However, little research has explored innovation deeply into routines' level. Second, how organizational routines evolve in an organization? Or how new organizational routines emerge? Existing research has offered abundant empirical support for endogenous change in organizational routines (Feldman 2000; Tsoukas and Chia 2002; Feldman and Pentland 2003), even in a setting such as invoice processing where such change should be unlikely (Pentland et al. 2011); little research has gone into exploring how routines change. Conducting an in-depth investigation into the case of Day-Definite (DD) Innovation in Arima World Group Company Limited (HOAU), this chapter attempts to uncover the black box of recreating organizational routines in implementing management innovation. Importantly, this chapter aims to address how the evolution of organizational routines promotes management innovation.

Following this introductory section, Sect. 5.2 of this chapter presents a literature review of the relationship between organizational routine and innovation, and components of organizational routines. Section 5.3 presents the method used. Section 5.4 show the case description and findings obtained respectively. Section 5.5 develops a framework and discusses implications, limitations, and future directions. Finally, Sect. 5.6 offers a conclusion.

5.2 Organizational Routine and Innovation

5.2.1 Organizational Routine

Since the concept of organizational routines was introduced by Stene (1940), it has been regarded as a central feature of human organizations and an explanatory mechanism in many of our most widely accepted theories, and as the primary means by which organizations accomplish much of what they do (March and Simon 1958; Cyert and March 1963; Thompson 1967; Nelson and Winter 1982).Given their importance, many scholars have theorized about the nature of organizational routines. Feldman and Pentland (2003) draw three dominant metaphors of organizational routines from existing extensive and diverse literature: programs, habits, and genes, and they argue that these metaphors treat organizational routines as mechanisms or abstractions, rather than as collective human activities. The focus of these metaphors is on the central tendencies rather than variation. As a result, these metaphors highlight the inertial qualities of routines and tend to minimize the possibility of flexibility and change. Therefore, looking deeply into the nature of organizational routines, Feldman and Pentland (2003) redefine organizational routines as repetitive, recognizable patterns of interdependent actions involving multiple actors. This definition highlights flexibility and change of organizational routines, which has been extensively accepted by recent researchers.

Traditionally, routines are seen as a source of stability in organizations and are designed to provide generalized solutions to recurring problems (Cyert and March 1963; March and Simon 1958; Ashforth and Fried 1988). Or, since organizational routines exhibit a great deal of continuity over time, traditional research even regard organizational routines as a source of inertia (Hannan and Freeman 1984), inflexibility (Weiss and Ilgen 1985; Gersick and Hackman 1990), and even mindlessness (Ashforth and Fried 1988). According to Geiger and Schröder (2014), the basic idea behind these conceptions is that organizations respond to recurring tasks in predefined ways. By allowing actors to respond to ex-ante defined tasks in ways that are also ex-ante defined, routines operate in a quasi-automatic manner and are therefore seen as the source of efficiency. Accordingly, routines are: "…repeated patterns of behavior that are bound by rules and customs and that do not change very much from one iteration to another" (Feldman 2000, p. 611). From this perspective, routines are routines because they do not change, but instead provide automatic and mindless responses to stimuli defined ex ante, and through their application lead to a predetermined result. Thus, routines ensure efficiency, legitimacy, accountability, and reliability in organizations. Although research highlights this functional aspect of routines, a related stream of research points to the potentially dark side of mindless and routinized behaviors.

Until 1982, Nelson and Winter first called attention to the centrality of organizational routines in their book *An Evolutionary Theory off Economic Change*, and henceforward, organizational routines have been among the core concepts in evolutionary economics, organization theory and strategic management (Becker 2004; Felin and Foss 2004). Following that, more and more recent research reveal that organizational routines are able to change continuously and endogenously, and thus see organizational routines as an important source of flexibility, change and innovation (Nelson and Winter 1982; Pentland and Rueter1994; Feldman 2000; Feldman and Pentland 2003, 2008; Feldman and Pentland 2005; Pentland et al. 2011, 2012; Bresman 2013). Research on routines has grown in recent years as scholars have increasingly recognized the centrality of this organizational phenomenon (Parmigiani and Howard-Grenville 2011; Salvato and Rerup 2011), which is devoted to routine dynamics, one branch of research on routines that is based in the idea that routines are practices with internal dynamics that contribute to both stability and change in organizations (Feldman and Pentland 2003). A core insight from research on routine dynamics is the close connection among routines, practices, and process (Howard-Grenville et al. 2016). Indeed, routine dynamics is based in the idea that routines not only connect inputs with outputs, but also that, as practices, they emerge through their own enactment and in relation to other practices (Feldman and Orlikowski 2011).

As studies have proliferated and scholars have added their insights, the term "routine dynamics" has come to stand for the study of the dynamics within and across routines as they are enacted in practice. Most research on routine dynamics has a common foundation of the articulation of performative and ostensive aspects of routines. Furthermore, Feldman et al. (2016) propose three core observations intrinsic to the work in routine dynamics: Action in routines is situated; Actors are knowledgeable and often reflective; What appears to be stable (for example, a routine) is only stable for now, at best.

More specifically, Cyert and March (1963) refer to the change of organizational routines as adaptation; while Nelson and Winter (1982) call it a mutation. Following these studies, Feldman (2000) claims that organizational routines have a great potential for change even though they are often perceived, even defined, as unchanging. Besides when there is a crisis (Gersick and Hackman 1990), in the early stages of establishing an organization (Narduzzo et al. 1997), or in areas of ambiguity (Miner 1990), routines can also change in old, established organizations in stable environments (Feldman 2000). For example, by modeling the routines as networks of action and using a first-order Markov model to test for stationarity, Pentland et al. (2011) find that routines generate hundreds of unique patterns that change significantly without any apparent external intervention, which offers empirical supports for an endogenous change in organizational routines. Further, Pentland et al. (2012) introduce a generative model of organizational routines to simulate and explain four important dynamic processes in routines: formation, inertia, endogenous change, and learning. Similarly, Loch et al. (2013) use an experiment with three-person groups to examine how routines evolve in the interaction of problem-solving and internal integration dynamics and find that groups do indeed develop problem-solving routines over time and use them consistently, and differentiated status within a group distorts its problem-solving routines by overweighting the influence of the high-status member. Miller et al. (2012) model the evolution of organizational routines as a process of formation, efficiency, and adaptability, and find that declarative memory built from past experience facilitates efficient routines in stable contexts, but obstructs efficiency gains when the organization encounters novel problems. Dittrich et al. (2016) examine the role of reflective talk in how routines change and argue that talk enables routine participants to collectively reflect on the routine and work out new ways of enacting it. All these studies confirm that organizational routines can change through a process affected by agents, actions, and memories.

5.2.2 Organizational Routine and Innovation

Traditional research insists that organizational routines have significant stability and are not easy to change. Organizational routines are the guarantee of remaining the stability or regularity and sustainable development of an organization, and can effectively ensure the organization accountability, political protection and conflict reducing; while simultaneously, organizational routines are also the root causes of organizational inertia, ignorance, skills reducing, negative motivations and ability traps (Hannan and Feldman 1984; March 1991). Organizational routines will lock organizations in inelastic, stable action paradigms so that organizations cannot adapt to the environment changes through innovations, thus organizational routines are the antitheses of innovations.

In recent years, on the contrary, research indicates that organizational routines have revolutionary or evolutionary (Pentland et al. 2010, 2011; Bresman 2013), mainly deriving from the implicit features of the participants. And researchers from both the organizational routine field and the innovation field have identified the relationship between organizational routines and innovation in recent decades. Studies on (management) innovation state that management innovation presents a systematic project potentially requiring fundamental changes in the routines of the organization (e.g. Argyris and Schän 1978; Birkinshaw et al. 2008) which lie on the most micro level of an organization and have been commonly understood to be a central element of organizations and a fundamental mechanism for coordinating work therein (Feldman and Pentland 2003; March and Simon 1958; Cyert and March 1963; Cohen and Klepper 1996; Ortmann 2010). McCabe (2002) particularly argues that management innovation is always intending to recreate routines of an organization. Meanwhile, this intention can be realized through implementation.

Recent research on organizational routines argue that organizational routines provide a basic definition of what an organizational change or an innovation really is at the organizational level, so they are regarded as the fundamental to understanding organizational change or innovation (Becker et al. 2005). Existing research on routines advocates a close relationship between organizational routines and innovation, which is based on the research of routine dynamics or routine self-change.

The idea that organizational routines can continuously and endogenously change leads researchers to suggest that routines are sources of organizational changes or innovations (Miner 1990; Feldman 2000). Existing research on innovation tends to remain much on a macro-level, which is incapable of capturing many interactions and their effects on actors and the environment. Feldman (2000) proposes a model of organizational routines to suggest that there is an internal dynamic to routines that can promote continuous changes and innovations. Or changes and innovations occur as a result of participants' reflections and reactions to outcomes of previous iterations of the routine (Feldman 2000). That is, focusing on routines enables researchers to explore into micro-level dynamics and identify driving forces of innovation on that level (Becker et al. 2005). Further, Daneshvar et al. (2012) identify external driving forces of routine evolution and confirm that these forces affect internal organizational routines and finally lead to organizational changes. Turner and Rindova (2012) set up a model to demonstrate the formation and coexistence of both stability and flexibility within an organization, and treat it as the source of organizational innovation. Using data on mergers and acquisitions affecting U.S. food and drug manufacturing plants, Anand et al. (2011) find that mergers appear to cause decay in adherence to routines, while acquisitions appear to serve as renewals and halt such decay. Heimeriks et al. (2012) also find successful mergers are able to form new high-order routines, resisting blocking effect from existing routines, to realize radical innovation after mergers. Accordingly, innovation can be realized through organizational routines.

In a word, the evolution of organizational routines in innovation implementation is the central issue that this chapter attempts to address.

5.2.3 Components of Organizational Routines

To explain the relationship with organizational changes or innovations, Feldman and Pentland (2003) suggest that, a deep understanding of organizational routines is needed. Feldman and Pentland (2003) propose two components of routines: the performative aspect and the ostensive aspect. The performative aspect "consists of specific actions, by specific people, in specific times and places. It is the routine in practice" (Feldman and Pentland 2003, p. 101). This aspect is directly observable. The ostensive aspect is "the ideal or schematic form of the routine. It is the abstract, generalized idea of the routine or the routine in principle" (Feldman and Pentland 2003, p. 101). The key distinguishing feature of the ostensive aspect is that it cannot be directly observed as it resides in the cognitive processes of the participants (Becker 2004). Some studies find the two aspects are recursively linked (Dionysiou and Tsoukas 2013; Feldman 2000; Feldman and Pentland 2003; Howard-Grenville 2005; Miller et al. 2012; Orlikowski 2002; Rerup and Feldman 2011); while others suggest that the ostensive aspect precedes and determines the performative aspect, that is to say, the ostensive aspect is on behalf of the practice of thought while the performative aspect enables the ideas presented (Feldman and Pentland 2005). The two aspects exist an interaction: some routines, formed under specific situations, will appear a variety of executive conditions, which reflects diversities and differences in the routine performing processes; while the result of performing routines will promote the changes of the ostensive aspect (Feldman and Pentland 2005; Rerup and Feldman 2011) and further make routines evolved, the changed performative aspect will further guide participants to complete next tasks (Dionysiou and Tsoukas 2013). From this we can see that organizational routines achieve their evolutions in the interaction between the performative aspect and the ostensive aspect, and further contribute to the realization of the organizational innovation.

It is clear that the performative aspect of an organizational routine is on behalf of the participants in the specific situation of specific actions (Feldman and Pentland 2003). Therefore, what needs to be further concerned is what the ostensive aspect consists of (Cohen and Bacdayan 1994; Lazaric 2008). Miller et al. (2012) find that the ostensive aspect of organizational routines consists of procedural, declarative and transactive memory and what individuals draw from their memories to perform routines. Dionysiou and Tsoukas (2013) use the Mead's concept of role taking to account for the fitting together of individual lines of action (performative) and the sharing of participants' schemata (included in the ostensive) as mutually constituted processes that occur as participants develop distinct selves in the context of a routine. They propose two key components of the ostensive aspect: shared understandings, also referred to as "shared schemata" (Becker 2004; Balogun and Johnson 2004; Feldman 2000; Howard-Grenville 2005; Tucker and Edmondson

2003), and a set of mutually coherent action dispositions called "coherent action disposition"(Dionysiou and Tsoukas 2013; Birnholtz et al. 2007; Hodgson 2008) which are retained in the form of procedural memory (Cohen and Bacdayan 1994; Feldman and Pentland 2003; Easterby-Smith and Lyles 2015). Shared schemata are individual schemata with intercoordination or complete identity, which enable the coordination of joint activities by supplying participants of routines with compatible interpretations about what is happening and reciprocal expectations for what is likely to happen next or what actions are appropriate (Blumer 1969, 2004; Cronin and Weingart 2007; Joas 1997; Miller 1973; Rentsch et al. 2008; Winter 2006); while action dispositions enable participants to respond to familiar inputs in an appropriate but also more unreflective way, thus economizing on the cognitive resources of participants (March and Simon 1958). Research further indicates that the creation of organizational routines relies on social interaction and role taking between participants of routines which enable participants to develop a joint understanding of the situation (Blumer 1969, 2004) and to align their actions appropriately (Mead 1934; Dionysiou and Tsoukas 2013). As Dionysiou and Tsoukas (2013) suggest, the (re)creation of organizational routines reflects a process of forming shared schemata and subsequent coherent action dispositions through interaction and role taking. That is to say, social interaction and role taking can help form shared schemata and coherent action dispositions, and integrate individual actions to form the collective paradigm, further make organizational routines evolved, and fundamentally achieve organizational innovation in the end.

In conclusion, research has explicated that organizational innovation was a complicated process affected by many factors, and implementing innovation refers to the combination of many elements and the coordination of each department. However, existing research has not been able to reveal the essence of innovation implementation and the key to success or failure. Following this research, based on the fact that the implementation of the organization innovation is to achieve the evolution of organizational routines (Chi et al. 2007), this research attempts to explore how organizational routines are recreated in a management innovation, i.e., the evolution of organizational routines with existing shared schemata and coherent action dispositions overturned and new shared schemata and coherent action dispositions set up.

5.3 Method

5.3.1 The Case Study

A longitudinal interpretive and exploratory case study is adopted here as a base for theory construction, which can find out the nature of the problem through the phenomena and build a theoretical model (Eisenhardt 1989, 1991; Yin 1994; Eisenhardt and Graebner 2007). On the one hand, although scholars explicated that the evolution of organizational routines was fundamental to achieve organizational innovation, little research has explored how organizational routines evolve in management innovation, hence, an exploratory case study is appropriate for developing a theoretical framework. On the other hand, the selected case here is a critical one that meets all the necessary conditions for developing an organizational-routine-evolution framework.

5.3.2 *The Case Subject*

The author selected Arima World Group Company Limited (HOAU), a fully-owned subsidiary of CITIC[1] Industrial Investment Fund Management Co., Ltd., as the sample case. Founded in 1995, HOAU was the first batch of companies awarded the China "5A" logistic company accreditation. On Oct. 26, 2010, HOAU was awarded "Top Ten Case Award of 2010 China Brand Building" with its industry-leading Day-Definite (DD) road transport service, as the only firm from the logistics industry. In 2015, HOAU was awarded "Asia Pacific Annual Award for the Best Brand Remodeling" by the world famous transformation magazine Awards. As a first rate transportation enterprise and the leader in the road transportation industry, HOAU has been working to build a strongest, fastest and most reliable distribution network, by offering Less-Than-Truckload (LTL) service, Full-Truckload (FTL) service,[2] and especially DD service.

HOAU experienced a long period of development. During the first ten years, benefiting from a short supply in the logistics industry, leading by the culture "arduous struggling, ahead forging, initiative changing and active competing", HOAU expanded from a ten-member store to the top three in the road transportation arena through integrating social resources to quickly occupy new markets. Now, HOAU has set up more than 3000 outlets nationwide. However, as the expansion of HOAU, its operating process had not changed along with the development of the company, industry, and even the change of customer demand, and therefore, the quality of its service was not improved, which led to a decline. Fortunately, one of the world's largest express delivery service providers Thomas National Transport (TNT) signed the Equity Transfer Agreement with HOAU Group in Shanghai in September 2006 to acquire its nationwide road transport and freight business.

[1]CITIC represents China International Trust and Investment Corporation which has developed into a large state-owned multinational conglomerate with both financial and non-financial businesses. By the end of 2014, CITIC's total assets stood at RMB 4732.9 billion, net assets RMB 267.6 billion, total revenue RMB 340.9 billion, and net profit RMB 29.1 billion. CITIC has been listed on Fortune Global 500 for seven consecutive years since 2009, ranking 186th in 2015.

[2]The LTL service moves goods from many different customers on one truck; while the FTL service moves full containers or trucks of one product from one customer. Therefore, the LTL service offers customers a more cost-effective method of shipping goods than the FTL service.HOAU offers more LTL service.

Hence, HOAU opened a new page of innovation and change under the direction of TNT. However, good times didn't last long and new problems always came with new environment. In 2008, the whole logistics industry was seriously attacked by the world financial crisis and suffered a lot from declining. The next year, HOAU introduced the concept of definite transportation from its international competitor, Debon Logistics, and launched DD innovation, a high-end highway express service product with the features of timing, security and high quality, which successfully freed HOAU from the dilemma and led the company to a new value-added arena, in terms of timing, security, and service quality. In order to achieve the high standard of DD innovation, it required a thorough change of the organization system, including a fundamental change of employees' cognitions. Hence, DD innovation was not only a product or service innovation but also a systematic management innovation. This research focused on it to explore the implementation mechanism of management innovation.

The preliminary proposal of DD innovation compassed the following three aspects.

First, HOAU started from the structure of the organization so that they set up a new DD Management Department subordinate to the headquarters to handle all DD innovation issues and appointed the vice president of Net and Process Optimization Department as the general manager. A DD product manager who directly reported to the vice president was to take charge of all issues within the department. Under the DD product manager, there were three team leaders, each responsible for product and service development, data analysis, product and project management respectively. More specifically, the product and service development team was responsible for examining new DD lines, setting up new networks, controlling operation quality, improving existing transporting lines, scheduling for new lines, analyzing KPI and other management indicators to improve it, and so on; the data analysis team was responsible for collecting all DD operation data, analyzing the data, finding problems and seeking solutions, providing data supports for other departments or teams, analyzing operation quality and gross profit, calculating DD proportion, examining satisfaction of customers and new-customer development, optimizing the weights of KPI, and so on; and the product and project management team was focusing on investigating market (referring to customers, competitors and the industrial environment), improving DD service in details, adjusting DD strategies, designing a standardized operation process of DD service, promoting DD service to new market, and so on. The DD management department took responsibility for all DD operation and management issues, while all other departments and management levels in HOAU would be engaged in implementing DD innovation, therefore, the new DD standard for the whole operation of HOAU put forward a new challenge, which led to the second aspect.

Second, HOAU repositioned DD service as the top-level service in highway express industry, so time nodes and the corresponding employee performance appraisal standards were set up in every link of DD service. HOAU set up new DD standards in terms of timing, security, and service quality. Compared with existing TLT service, DD service was of more reliability and higher service quality, for example, definite one day arrival but only a 25% higher price; while compared with air express service, DD service was of equivalent quality, but charged only one third of the price. The target customers of DD service were medium and small firms from high value added industries, such as automobile parts, computer, telecommunications and electronics, pharmaceuticals, and instrument and meter industry. They paid most attention to goods security, and then the speed and price especially focusing on the goods status tracking service when choosing logistics providers. Aiming to ensure the punctual arrival of DD goods through the control system monitoring the whole process, high standards and strict requirements for every transport node, the set-up standards for timing consisted of controlling rules for all points in the transportation process, such as, forbidding any over-one-minute delay of departure (the industrial standard was 30 min), loading or unloading goods within six hours, installing all trucks with GPS to make sure a 24-h controlling, and picking up or delivering goods as promised. In order to ensure the security of DD goods, a set of standards were set up in the links where the security problems were easy to appear (for example, warehouse, loading and unloading, transportation, etc.). The standards for security referred to installing the DC system, through which HOAU could count and sort the data of daily operating conditions, analyze the operation problems, then transmit orders to branch headquarters so that branch headquarters could adjust the work errors; establishing a special DD stocking area, which provided a special place to store DD goods separated from others, achieving the goal of controlling error rate under 0.08% (far superior to the industrial average); packaging DD goods with special steel cages in order to transport DD goods by completely enclosed trucks; taking delivery of goods with delivery password and the consignee's valid certificate; making special marks for high-value goods and so on. Finally, HOAU set up high standards for service quality including establishing a DD sales team, giving prior to DD goods at any time, setting up a DD customer service team and a DD hotline, and offering customized service, all of which aimed to ensure that customers could experience the high standards of exclusive services of DD service at any place.

Third, HOAU recreated a DD business process and clearly defined each link. As shown in Fig. 5.1, after picking up or receiving DD goods, HOAU stores first sorted and packaged them, and labeled them with letter "D". At the same time, HOAU stores put the information on these goods into DC (the DD information system) and then delivered them to the Transfer Center. The Transfer Center resorted and repackaged the goods with special steel cages, scanned and then loaded them into DD trucks for completely-enclosed transportation. The Dispatching Center dispatched trucks through DC system, controlled the main-line transportation through GPS and gave orders in an accident, to make sure goods arrive at the Destination Transfer Center within 48 h. The Destination Transfer Center then unloaded goods, checked the information and the package, put exceptions into DC if any, and delivered goods to destination stores. Finally, employees from stores further checked the package, put exceptions into DC if any, and delivered them to receivers. If receivers were not satisfied with the delivery, stores would take responsibility to negotiate with them and sign a contract on

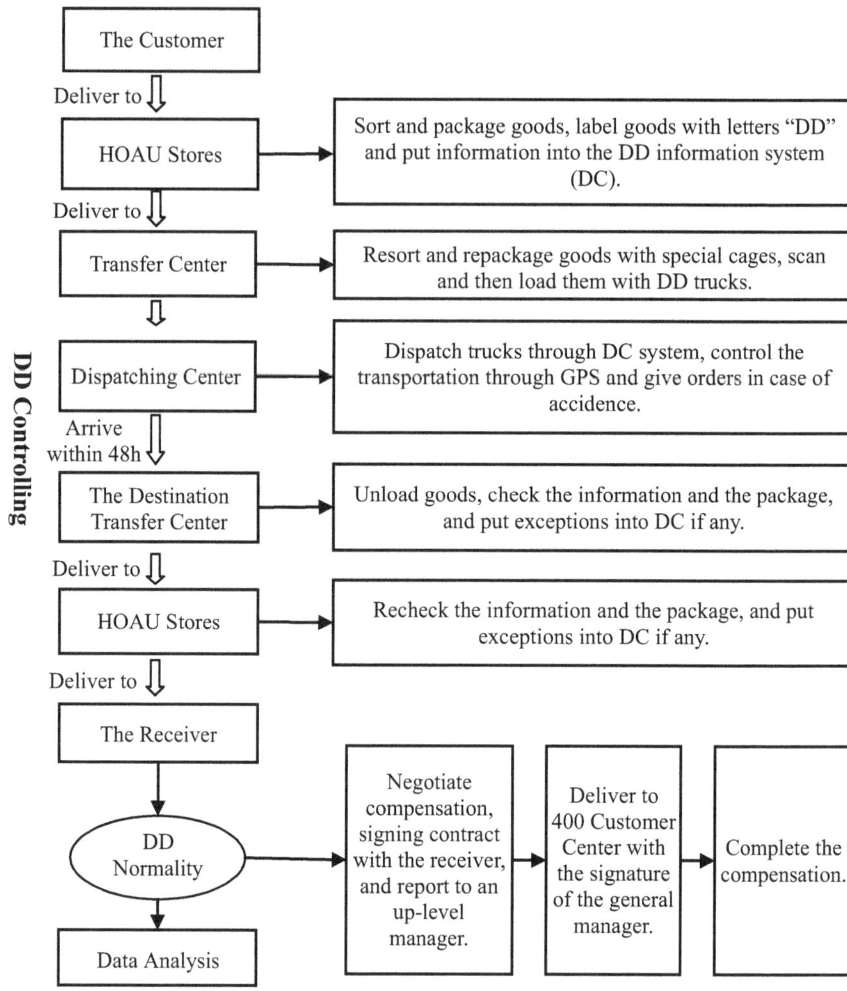

Fig. 5.1 The DD business process

compensation, then report to an up-level manager for checking, deliver to 400 Customer Center with the signature of the general manager, and finally completed the compensation.

5.3.3 *Data Collection*

Data collection took place from Jun 2015 to Sep, 2015. This research utilized multiple approaches during data collection to meet criteria for trustworthiness

Table 5.1 Quantitative details of interview data

Informant	Jun 12, 2015, and Jun 16, 2015	Jul 20, 2015–Jul 25, 2015	Jul 30, 2015–Aug 10, 2015
Members of DD Management Department, including Vice President of Net and Process Optimization Department(also the department manager), DD Product Manager, Directors of Product and Service Development, Data Analysis, and product and project management	1* + 1^		
Branch headquarters of Zhengzhou, including General Manager of Zhengzhou Branch, General Office Director, Data Operation Director, and several employees from these departments		1* + 2^	
Managers and Employees from different phases of DD Process, including No. 7 Store under Zhengzhou Branch, Transport Department, Sales Department, Transfer Center, and Station Dispatching Group			2* + 22
Total	2	3	24

Note A "^" indicates a telephone interview. A "*" indicates a group interview

(Lincoln and Guba 1985; Yin 1994), including semi-structured interviews, archival data, and observation. Semi-structured interviews offered most of the information concerning how HOAU implemented DD innovation and how organizational routines evolved. Archival materials and naturalistic observation offered background on the firm and expanded the understanding by offering insights which clarified interview findings.

Semi-structured interviews. Since this research focused on the implementation of the innovation process, rather than the early stages of the decision of innovation, this research put the view on investigating the DD Management Department responsible for DD matters located in HOAU headquarters, the Zhengzhou Branch and some stores whose implementation effects were satisfied. Consequently, as summarized in Table 5.1, a total of 3 sets of 29 interviews were conducted and transcribed, including 22 individual interviews, 3 telephone interviews, and 4 group interviews. The first set of 1 group interview with members from DD Management Department, referring to Vice President of Net and Process Optimization Department (also the department manager), DD Product Manager, Directors of Service, Data Analysis, and Product Research and Project, and 1 telephone interview with the Vice President, was conducted on Jun 12, 2015 and Jun 16, 2015 respectively. This research focused on gaining general information on HOAU and DD innovation, including the development process of HOAU, the background of DD innovation, innovation content, implementing the process, existing problems and solutions, and innovation results. A descriptive case on DD innovation was written based on the data from this set of interviews (see Sect. 5.3.2). The second set was conducted during Jul 20, 2015 and Jul 25, 2015, including 1 group interview with members from Zhengzhou Branch Headquarters (including General Manager of Zhengzhou Branch, General Office Director, Data Operation Director, and several employees from these departments), and 2 telephone interviews with General Office Director and Data Operation Director. This set of interviews focused on details of how DD innovation was implemented in Zhengzhou branch and what effects the innovation produced. Representative participants involving in DD system were chosen by the general manager from each link for further exploration. Finally, the third set of interviews consisting of 2 group interviews with members of No. 7 Store and Station Dispatching Group of Zhengzhou Branch and 22 individual interviews with key individuals from different links (for example, No. 7 Store under Zhengzhou Branch, Transport Department, Sales Department, Transfer Center, and Station Dispatching Group) was conducted during Jul 30, 2015 and Aug 10, 2015. It aimed to confirm how cognition of routine participants evolved during the implementation process of innovation.

In view of the inductive aims, the author and the research team members encouraged the informants to go toughly into details. For each set of interviews, an interview protocol was designed with major themes in mind; and during the interviews, questions were not asked in any specific order but were organized by the actual situation of the informants (Gummesson 2000) and some additional problems might be proposed to deeply excavate more valuable information and find out some hidden details.

Archival data. This research supplemented interviews with over one hundred documents containing HOAU introduction, services introduction, DD process diagram, DD daily reports, DD quality reports, operation data of Zhengzhou branch, meeting notes, memos, and annual reports. These documents were collected not only to track the development of HOAU, the process of DD service, innovation activities, tasks for individuals, and the interaction between links in details, but also to provide supports for the implementation of organization innovation and the evolvement of organization routines. Some documents were downloaded from the OA (Office Automation) system of HOAU, from June to August 2015 when one of our research team members was working as an internship in the Data Operation Department of Zhengzhou branch; while others were offered by the institution office. In addition, the author gathered articles, media reports, stories and Web materials about HOAU and the transportation and logistics industry. The author created categories for filing, retrieving, and analyzing the archival data, to support interviews and the findings.

Observations. During each visit, the author took notes of the informal observations gaining the information of the actual operation processes such as DD operating system and dispatching DD goods while waiting for interviews and walking around the firm. The dining hall and the dormitory offered the opportunity to have free talks with employees and observe their interaction ways and contents. Particularly, when visiting No. 7 Store, the author got the opportunity to observe how employees loaded and unloaded DD goods as well as how to scan and take notes of DD goods, and then made a comparison with that of TLT goods. The research member of the team also made specific notes about how DD system was operated, how employees interacted with each other and how DD tasks were finished.

5.3.4 Data Analysis and Coding

The process of analysis was guided by Eisenhardt's notion that "it is the intimate connection with empirical reality that permits the development of a testable, relevant, valid theory" (1989, p. 532). Throughout analysis and coding, techniques such as constant comparison (Glaser and Strauss 1967) and content analysis (Krippendorff 2004a, b) were used because "no analysis strategy will produce theory without an uncodifiable creative leap, however small" (Langley 1999, p. 691). This not only enhanced data interpretation, but also increased confidence in the process (Golden-Biddle and Locke 2007).

Considering the scholars' theoretical guidance for the analysis process and the actual situation of this case study, the author divided the data analysis process into five steps. First, data was collated and sorted, with raw data transformed by recording and transcribing interviews, collating field notes, and writing up observations. As the primary source, interviews offered the most information on how HOAU implemented DD innovation and how participants' cognition and behaviors changed during the innovation process as well as how organizational routines evolved. Second, based on collected information, a descriptive case about DD innovation was written to show the main content of the innovation and to extract the relevant content of the organizational structures, processes and standards (see Sect. 5.3.2). In the third stage, via blending data and sequential analysis, a story of DD innovation implementation was written to capture the flow of innovation activities and evolution of organizational routines. And the implementation process of organizational innovation was divided into three phases: the existing-organizational-routine-domination phase, the new-routine-creation phase and the new-routine-solidification phase. Then, the author adhered to guidelines specified for methods of naturalistic inquiry and constant comparison techniques (Lincoln and Guba 1985; Strauss and Corbin 1990) to analyze how existing shared schemata and relevant coherent action dispositions (two components of the ostensive aspect of organizational routine) were overturned and then recreated by focusing on two aspects: innovation activities taken by HOAU, and changes of participants' cognition and behaviors. Finally, a framework showing how organizational routines evolved in innovation was set up, with three phases of existing routine domination, new routine creation and new routine solidification.

Systematic, iterative comparisons of data, emerging categories, and existing literature aided the development of cohesive constructs and an integrative, theoretical framework (Andriopoulos and Lewis 2009). As the author discerned codes that were similar, the author collated innovation activities or the direct descriptions indicating similar meaning into first-order activities or categories. Then, the author discerned linkages among the categories that could lead to the development of second-order themes by formulating researcher-induced concepts at a more abstract level. Finally, the author assembled the second-order themes of innovation activities and participants' cognition and behaviors into aggregate phases to develop a process framework.

5.4 Case Analysis: Evolution of Organizational Routines in Innovation Implementation

According to the innovation proposal (see Sect. 5.3.2), DD innovation of HOAU was a systematical project referring to organizational structure change, market reposition, standards reset up, business process redesign, and operation model reestablishment. However, the establishment of the proposal was only the beginning of innovation. It was the implementation that transferred value into practices and produced innovation effects. Actually, it took HOAU almost three years to implement DD innovation before a final success. During the implementation period, changes took place not only in organizational structures, service standards, service processes and other aspects of the dominant changes, but also in organization routines which were on behalf of organizational attribute characteristics and employee cognition. The existing organizational routines reflecting fundamental features of HOAU and employees' cognition and behaviors were overturned and new organizational routines for DD service were created. From an explicit perspective, the implementation exhibited a series of continuous activities taken by HOAU; while from an implicit perspective, it reflected gradual changes occurring in DD participants' cognition and behaviors, namely the evolution of organizational routines. In previous studies, scholars mainly focused on organizational innovation activities, while ignoring the micro-level of participants' cognition and behavior changes. Therefore, this research attempted to explore the evolution mechanism of organizational routines in innovation implementation by considering both explicit innovation activities and implicit changes of participants' cognition and behaviors. A three-phase process of evolution was identified, including the existing-organizational-routine-domination phase, the new-routine-creation phase and the new-routine-solidification phase.

5.4.1 The Existing-Organizational-Routine-Domination Phase

Before DD innovation, HOAU had been focusing on offering LTL Service since its inception. LTL Service relied on large volume and low prices in the competition. Therefore, the vision of HOAU at that time was making profits through enlarging market coverage and increasing transport volume; and consequently, the income of employees depended on bonuses based on the volume. In addition, each link in the process of LTL service was independent. Each finished their tasks and got appraised independently. Members of HOAU described LTL Service as following:

> At the time of LTL Service, every year, different operation regions of HOAU competed with each other in total volume for bonuses. At the same time, the competition in the logistics industry intensified year by year. So our goal was to expand HOAU by setting up more stores and attracting more customers, and thus to increase the volume and defeat rivals in speed. (Manager of Zhengzhou Branch, Jul 20, 2015)

> Volume was our unique target when offering LTL Service. All our members, no matter salesmen, loaders, unloaders, or other workers, tried their best to perform more tasks, so as to get more bonuses for individuals and also for the whole store. (Manager of No. 7 Store, Jul 31, 2015)

> At that time, the individual loading or unloading volume was several times that of nowadays. We climbed up to the top of trucks and directly threw goods down. Of course, we tried to avoid our goods being seriously destroyed. (An employee from No. 7 Store, Aug 2, 2015)

Accordingly, the organizational routines based on LTL Service can be concluded as the action patterns focusing on volume maximization and independent tasks. That is, during the period of LTL Service, the common understanding of all HOAU's members towards the organizational task was volume maximization, and the coherent action dispositions reflected the tendency of performing individual tasks as much as possible and as quickly as possible. The ostensive aspect of organizational routines focused on volume and speed, rather than service quality or service details.

Innovation activities for preliminary implementation. DD innovation started as scheduled in February, 2009. HOAU designed a series of innovative initiatives and formed the innovation system. The implementation of DD innovation began with setting up DD management department, assigning the manager and three directors, and arranging specific tasks for them. Meanwhile, HOAU proved the reliability of the innovation proposal through a test conducted in new stores. After that, HOAU proved that DD innovation was effective and could be spread to the majority of their branches and stores. Taking this into consideration, HOAU selected qualified branches and stores to run DD service to support the traditional LTL business and implement DD innovation over round the whole firm, which was the real beginning of DD innovation. To do so, HOAU redivided its nine districts (including the Northeast China, North China, Central Plains, East China, Central South, Zhejiang Min, Shandong, Western China and South China) into 56 branches based on DD requirements and costs, including operation districts and business districts. The operation districts were responsible for goods transportation; while the business districts were for value controlling and standardization. HOAU also rescheduled DD transportation lines by transforming city-to-city single lines to main lines. For example, before rescheduling, there were several city-to-city lines between the South District and the Shandong District, each connecting one city from the South District and another from the Shandong District; while after rescheduling, there were only two main lines, one connecting Jinan city and Guangzhou city, and the other connecting Qingdao city and Guangzhou city. DD goods from other cities were gathered in these four cities for further delivery through the main lines.

The optimization of network layout and the reducing of the number of transfers not only reduced the cargo damage rate of the DD product, but also improved the security of the DD goods. Each of DD goods would be loaded into the exclusive cages, so it was easy for employees to use the scanner to better tally DD goods leading to fewer errors.

Moreover, HOAU purchased new trucks with Global Positioning System (GPS), and set up Office Automation System (OA) and a new DD Information System (IS) for DD service. OA and IS were applied in many areas, such as dispatching DD goods, preparing for transportation, controlling transportations, and more importantly, collecting data on details of DD operation. The DD management department analyzed these data every day to find out problems and errors in DD operation, formed reports and delivered results down to the headquarters of HOAU branches. According to the daily analysis results, the branch headquarters adjusted their activities to reduce operation errors and damages as well as control the Rate of Error and the Rate of Damage under 0.08% and 0.8% respectively. Additionally, HOAU branches utilized IS to allocate trucks, assign drivers, and check transportation times and status; Stores relied on IS to input orders, develop customers, deliver goods in time and guarantee the safety of loading and unloading. In the early stages of full implementation of DD innovation and these innovation activities helped HOAU successfully attract abundant of new DD customers owing to high service standards. However, at the same time, HOAU lost lots of existing customers who paid the high price of DD service but received equivalent service as LTL Service. This was because new DD standards were not really put into practices. When offering DD service, most HOAU branches and stores still held existing standards of LTL Service.

Cognition and behaviors of participants. At the beginning, most new DD standards set up by HOAU were ambiguous. Implementation of these new but ambiguous standards or innovation measures brought all employees engaging in DD innovation (namely participants of DD innovation) with perceived uncertainty which led to anxiety and loss of security. In order to reduce anxiety and insecurity, participants strived to turn novelty into familiarity through analogical transfer which helped them seeking for the similarity of the new context and the existing context by mapping categories and relations from the existing context to a new one. That is, in the early period of implementation, participants of DD innovation tended to intentionally keep their attention on the similarity of DD service and LTL Service to make their tasks continuously performed, but neglecting the difference. They hesitated to transfer their attention to new DD service for anxiety and loss of security. Participants of DD innovation described their cognitions on DD innovation at the beginning of the implementation as following:

> When the Shanghai headquarters announced to open DD service in our branch, to be honestly, we were shocked. The DD management department offered us with new DD standards, showed us the DD business process, shared us with successful experience, and confirmed a bright future of conducting DD innovation. Nevertheless, we were anxious about the uncertainty of innovation. Moreover, we knew that the whole branch would be overburdened if we followed all new standards. We had been offering LTL Service for a long period of time and the LTL thought had rooted in our mind. At the beginning, there was no clear boundary between new DD service and existing LTL Service, because our employees were not willing to accept the difference. We mixed them together by offering them in a similar way or even the same way. Actually, most internal rules or institutions still support us to behavior in an LTL way. That was also a reason why our employees did not want to make changes. (Manager of Zhengzhou Branch, Jul 20, 2015)

> We suffered a lot from anxiety and insecurity at the beginning. Soon, we realized that it was too difficult to follow all new standards of DD service, because of limited employees and lack of experience. Moreover, the volume of DD goods was not stable. Sometimes there was too much to been loaded; while some other times, there was not enough for a full truck. All that we could do was to make DD service a little better than LTL Service. (Manager of Zhengzhou distribution center, Aug 4, 2015)

> DD innovation set up a series of new requirements for goods packaging. However, in the mind of our employees, little difference existed between packaging of DD goods and that of others. What they actually did was adding more nails, or more shock and protection layers. After all, the volume was the first and most important goal. (Manager of No. 7 Store, Jul 31, 2015)

> I was really shocked by new requirements of DD service, and I did not think most of these requirements were necessary (at the beginning). Before that, I had been honored as 'Outstanding Individual' for several years for my high working efficiency. I was familiar with loading, unloading, packaging and delivering. Also, I did not want to slow down to handle DD goods in a new way and lower my income. When I got a complaint from a DD customer because of a late delivery for half an hour, I was really angry. I did not mean to do that. It was an accident. (An employee from No. 7 Store, Aug 2, 2015)

Accordingly, during this early period of DD innovation implementation, participants were unwilling to engage in the innovation because of the perceived uncertainty stemming from ambiguous innovation proposals and goals. They strived to keep existing routines and follow existing rules and regulations in continuously performing their tasks. Further, rules and regulations of HOAU, such as the store management regulations, the assessment system, the employee management regulations, and the quality management regulations, were not reestablished, which were judged by participants as appropriate for the situation they face. Therefore, existing organizational routines based on LTL Service dominated participants' cognition and subsequent actions in this period. In a word, during this existing-organizational-routine-domination phase, due to the ambiguity of the standards and innovation processes and consequently the insecurity of the participants, they preferred to rely on analogical transfer to search for the similarity between the new and existing context, emphasize the appropriation of existing rules and regulations appropriate for new situation, try to maintain the existing cognitive and behavioral paradigm and take similar action, so the existing organizational routine was still dominant in the practice of guiding and standardizing the behavior of employees. That is, by emphasizing the similarity of the new task and the existing tasks and the adaptability of the existing system, most of the uncertainty brought by DD innovation could be absorbed, thus the participants were still able to complete the task in an old way.

5.4.2 The New-Routine-Creation Phase

With the standard and the procedure of DD innovation and each link in the process of DD innovation becoming more and clearer, the participants realized that holding existing cognition and behavior patterns became more and more inappropriate for performing tasks. Meanwhile managers of HOAU also realized that implementation of DD innovation was not only a simple process of putting new standards and process settings into practices, but also a complicated process of transforming cognition and behavior patterns of HOAU and establishing new routines. Therefore, HOAU continued to take innovation activities which were accompanied by change of participants' cognition and behaviors.

Innovation activities for deep implementation. First, HOAU set up a new learning and training system for delivery of innovation missions and requirements for all employees, especially for first-line employees who relatively lacked of knowledge but directly operated the DD innovation. The DD management department designed videos, booklets and presentations for purpose of DD innovation, characteristics of new services, new standards and processes, key points in implementation, and its contribution to future development of HOAU and individuals, and then exhibited them through the set-up learning system. This facilitates all levels of HOAU, no matter stores, centers, branches or departments of the headquarters, freely organizing learning activities. Moreover, the concept and the mission of DD innovation were also delivered to employees through collective learning. First-line employees, also the operators of DD innovation, whose education levels were relatively low, were listed as an important learning objects. Managers realized that employees can change their cognition only when DD culture arrived at the operation level. Another activity of learning and training focused on how to operate IS system. IS system was a new logistic recording and controlling method and also a key technology for guaranteeing the high quality DD service and timing. When facing with new things, different people had different degrees of acceptance, for instance, some old staff limited by the level of culture tended to object to new technologies. To solve this problem, HOAU organized a professional team to teach all employees in firm that how to use IS hand by hand, helping employees benefit from its convenience and accuracy and shorten their time for getting used to it. Took an unloading employee for example, after he learned how to use a PDA scanner, it was difficult to curb his excitement. He said: "before these trainings, all of us always needed to remember a bunch of waybill numbers and goods numbers, but now we could use this scanner to put a lot of data into the computer system". Further, HOAU offered employees with special trainings on reception, marketing and hot line answering. Specifically, the training on reception focused on how to select proper terms for customers with different prices and delivery demands; the training on marketing referred to how to set up customer development schedule and how to make follow-up calls; and the training on hot line answering reflected skills of proposing best answers for DD customers on hot line. All these training activities helped employees to get deeper understanding on DD innovation.

From the perspective of the company's point of view, all these training activities contributed to promote the efficiency and quality of DD innovation; from the employee's point of view, these helped employees to get deeper understanding on DD innovation and learn how to work better. However, just knowing how to do was not enough, employees needed to have a positive job initiative and enthusiasm. With the same number of employees, stores and branch offices were asked to complete LTL Service and DD innovation at the same time. In fact, the workload of the staff greatly increased. Since the DD service quality was not the latest, and the system by which the company allocated the bonus according to employee's performance could hardly come true, this resulted in a vicious circle that the employee's work was greatly increased but their income didn't increase, leading to the rejection to the DD innovation from core staffs.

So, HOAU connected participants' individual interests to DD interest and encouraged them to identify new roles in new situation. For example, HOAU adjusted the performance appraisal and bonus system which had a direct impact on employee motivation. The new bonus calculation formula was "Bonus = (Individual Bonus Base * Value Discount * ABCD Coefficient + Non vehicle overcapacity Bonus) * DD Coefficient + Arrival Bonus + Complaint Penalty + Full Vehicle Bonus + Rising Feet and Weight Penalty". In the new formula, DD Coefficient was added. When DD goods were not handled properly, all links in the processes would be interrupted, which was different from LTL Service process. In LTL Service process, tasks of each link were independently undertaken and appraised; while in DD process, all links were united together, sharing the same DD Coefficient. This new performance appraisal system emphasizes the overall interest of the whole DD process involving all participants.

Cognition and behaviors of participants. Repeated experiences in carrying out DD tasks and more innovative activities facilitated recognition of distinctions between DD Service and LTL Service, rather than their similarities. Meanwhile, participants identified the inadaptability of existing rules and regulations. Changes of cognition increased perceived uncertainty of participants which forced them to seek interactions with other participants especially those from linking stages who had closed relationships. Through interactions, they were able to gain more information about DD service, to make their tasks continuously performed, and finally to find their new roles in new tasks. This is the beginning of existing organizational diminishing routines and new organizational creating routines.

Since each link was independent in LTL Service, participants from different links did not know each other, let alone their roles or cognition. In order to exchange information and judge position of their own, they actively and frequently communicated despite their different responsible areas, especially concerning how to perform tasks. Through the interaction pattern characterized by high frequency explicit communications, participants were able to identify roles and contributions of others and form a joint cognition in DD innovation.

For example, in No. 7 Store, a salesman stopped a savage unloading by saying that "They are DD goods. Please stop unloading in this way anymore. Otherwise, how can I introduce DD service to our customers?" The store manager added: "Now, speed is not the unique criteria. We should pay more attention to the quality. Guarantee efficiency and quality of every stage, so that we can improve our performances and benefits from DD innovation."

Another example reflecting the interaction happened between employees from No. 7 Store and a truck driver. When DD goods arrived at the store, due to bad weather the expected delivery time was approaching. The driver participated in unloading and checking activities without hesitation, rather than leaving immediately as before. Moreover, he repeatedly reminded employees taking care of the goods, since "they are vulnerable electronic goods with high value". Why the driver knew clearly about the goods was because of an interaction between him and the pervious link.

Explicit communications and interactions between linking stages frequently occurred in performing DD tasks, which was different from LTL Service time. Actually, explicit communications and interactions were the origin of new routines. Through high frequent explicit interactions, most of which were face-to-face communications, participants gradually took into account roles of others especially those from linking stages, including their understandings, ideas, opinions, and consequent actions in performing the new and joint activity of DD service, which indicated participants engaged in role taking. Consequently, they were able to develop a joint, situated understanding of the concrete situation, identify appropriate actions, and align their individual lines of action accordingly, first between linking stages and gradually among all stages. Role taking made individual participants pay closer attention to joint activities rather than individual tasks, and see himself or herself as a part of joint activities or a social self rather than a single self. Moreover, participants relied on role taking to understand the intentions of others, predict their actions, and make corresponding responses; subsequently they began fitting individual lines of action together to form joint action. The joint understanding was the outcome of interaction and was inseparable from the context of DD innovation. Participants used the joint understanding and effective expectations to guide their own actions. For example, loading employees were able to predict the understandings and actions of participants from the previous stage (i.e., packaging employees) and these from the next stage (i.e., drivers); and the store manager was able to predict the reaction of site scheduling manager when DD goods did not arrive in time.

The development of joint understandings through role taking was a critical step towards the development of each participant's "self" in the new DD process. Development of their roles and contributions to DD service through abstraction and generalization can be called individual schemata for DD tasks. Their individual schemata referred to several knowledge domains, such as roles, situations, and events and their sequences and their actions. Participants abstracted and generalized from both personal understandings and actions and joint ones that had been established with fellow participants through interactions that further helped to develop a self-belonging to joint activity of DD service. Since each participant abstracted and generalized from situated meanings, individual schemata had been developed to become at least partially shared among participants, especially between linking stages. Schemata in turn helped individual participants to develop some sense of prediction and controls in performing DD tasks. They were able to organize their behaviors or actions in accordance with mutually consistent expectations of how they should act in certain situations for performing DD tasks. This facilitated and guided the fittings of individual lines of actions in future activities and increased the efficiency of communication and coordination among participants. However, participants paid more attention to those having close relationships with them or having more effects on them and engaged in role taking, such as the interaction between site scheduling workers and drivers, drivers and unloading workers, and unloading workers and checkers.

As participants formed common understandings on DD service and the requirements of high quality and timing, shared schemata for the joint activity emerged between relational participants and finally among all participants. Shared schemata enabled participants to anticipate understandings and behaviors of others, establish confidence about how to behave, and meet basic requirements for the coordination of joint activities. Also, shared schemata supplied participants with mutually consistent interpretations and evaluations of information, as well as with reciprocal expectations concerning what actions were appropriate for the situation they should face. As an essential component of the ostensive aspect of new organizational routines for DD service, shared schemata facilitated realization of compatible, reciprocal behavioral expectations. Each participant was able to correctly expect that he or she would receive familiar signals from the others and would respond in familiar ways, even without explicit communications. That is, the frequency of explicit communications between relational participants gradually declined; while the effect of role taking increased.

Shared schemata enabled participants to further establish habits and skills in packing, delivering, recording, dispatching and recovering and develop coherent action dispositions towards timing, security and high quality. Action dispositions, which were retained in the form of procedural memory and were included as elements of the ostensive aspect of the routine, made participants respond to familiar inputs in an appropriate but also more unreflective way, thus economizing on the cognitive resources of participants. For example, holding the common understanding that serious damage controlling was a criterion of DD service, participants from each stage paid close attention to it. When a serious damage on a truck of fragile products was caused by accident in July 2012, all participants did their best to make it up. They saw themselves as an indispensible part of DD process, with a set of cohesive action dispositions towards the same organizational goals. The creation of shared schemata and subsequent action dispositions reflected

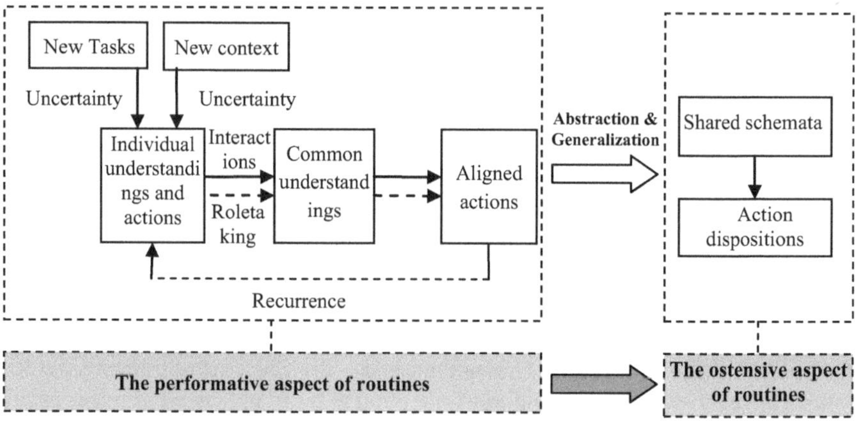

Fig. 5.2 The creation of new routines in innovation

the underlying change of participants' cognition and behaviors, and also emergence of the ostensive aspect of routines.

In conclusion, during this new organizational routine creation phase, as DD tasks and the new context become clearer and clearer, and more innovation activities were taken by HOAU for deeper implementation, participants realized that holding existing cognition and behavior patterns were not an efficient way of meeting DD requirements and that the existing rules and regulations were not able to absorb perceived uncertainty. Then, they transferred their attention from the similarity of the new context and the existing context to their distinction. In order to reduce perceived uncertainty and anxiety, participants began to seek for interactions with other participants especially with those relational ones. The interaction helped participants find their new roles in joint activity of new tasks, develop new impression of cognition on understandings, ideas, and thoughts of others, and effectively predict their actions when facing certain situations. This role taking facilitated creation of shared schemata among participants and subsequent action dispositions, which indicated the creation of the ostensive aspect of new routines. The creation of new organizational routines in innovation is showed in Fig. 5.2.

5.4.3 The New-Routine-Solidification Phase

After the emergence of new routines for DD service, HOAU took more activities to strengthen them and gradually weaken the impact of LTL service. At the same time, cognition and behaviors of participants continued to evolve.

Innovation activities of HOAU. HOAU disseminated information about successful DD examples to all over the firm to attract more attention on DD innovation, which helped solidifying new routines and reducing the effects of previous ones at the same time. Examples included how No. 6 Store in Zhengzhou city finally met all criterions of DD service and improved its performance when facing the disadvantage of remote location, low employee quality and low customer satisfaction; and how the main line between Qingdao and Dongguan kept the record of zero complaint for a whole quarter. These real examples gave participants positive information that could effectively motivate work enthusiasm and reduce uncertainty. In the actual operation of enterprises, carrots and sticks might be a better way to motivate participants. Also, HOAU demoted a store manager for failure to meet DD criterions for 3 consecutive months, reminding participants the importance of liberating from previous routines. More importantly, when realized that new shared schemata and action dispositions were created, HOAU issued new rules and regulations on DD criterions, DD process, compensation and rewards, appraisement and management, based on a collection and analysis of comments and suggestions from participants. More specifically, HOAU designed DD flow charts and showed them in the center of every store and branch; clarified criterions of DD service in terms of quality, efficiency, costs, timing and security (for example, criterions of quality reflected error rate, abnormal sign rate, goods lost, goods shortage, complaint rate, and scanning rate; criterions of efficiency referred to self-owned truck preparing efficiency, self owned truck attendance, average volume, and employee attendance; criterions of cost included per ton cost, Kilometers maintenance cost, and full load rate; criterions of timing referred to dispatching punctuality rate, delay rate, unloading punctuality rate, transforming punctuality rate, delivering punctuality rate, delivery success rate; criterions of security referred to million tons of work-related injury rate and one hundred thousand km accident rate); reset up the quality management system to focus on DD criterions and specialty of DD goods; revised the performance assessment systems, including the department level assessment system, the district level assessment system, the line area assessment system, and the store assessment system; and improved the calculation formula of DD coefficient. All these newly set-up rules and regulations became the carrier of new organizational routines.

Cognition and behaviors of participants. As new organizational routines with common understandings and cohesive action dispositions emerged and became more organized and elaborate, they supplied participants with more guidance for their subsequent conduct in performing new tasks. Participants transformed their cognition towards specialty of DD service and aligned their actions with new rules and regulations. Besides explicit changes in rules and regulations, participants internally and initially established new rules for new tasks in repeated experience by creating new ways, techniques and know-how, and revising or discarding existing ones. These endogenous rules for DD service increased the likelihood that participants would find appropriate inferred solutions from past performances to deal with new problems. Both exogenous and endogenous rules progressively absorbed an increasing part of perceived uncertainty so that, over time, a diminishing amount of uncertainty remained to be absorbed through role taking and the realignment of individual schemata. Participants described the change as the following:

> After a period of implementation, DD service became the dominating business in our firm in 2012. A complete operation system of DD service was set up. This opened a new page of development for HOAU. Now over 60% of our profit is from DD service. What's more important is that the DD mission, criterions and requirements have deeply rooted in the mind of participants, no matter from stores, centers, lines, or the headquarters. The DD process unites all participants together to work for a common goal. We can see that the service quality has been radically improved, including that of LTL service. Moreover, we have set up rules and regulations to support new tasks, and have created our way of dealing with DD issues. (Manager of Zhengzhou Branch, Jul 20, 2015)

> It took three years for us to transform their mind from LTL service to DD service (the process of creating new routines). The company had provided us with a lot of supports from top to bottom, including the equipment of new vehicles, the systematical training of all employees, and the adjustment of bonus system, after which employees worked harder and harder. We have mastered skills of DD scrvicc. Now, we are able to identify any detail or any tiny problem in checking, scanning, or packaging, and appropriately handle them. (Manager of Zhengzhou distribution center, Aug 4, 2015)

> We experienced a difficult time when HOAU started DD innovation, but everything became different when we got ourselves completely involved in the new tasks and changed our mind. We set up a separate packaging center, a loading and unloading center, and a scanning and storing center for DD goods, to guarantee the quality. Now, though our workers perform both DD and other tasks, they know perfect well the specialty of DD service. In the past, they did not want to spend any time waiting for our customers or waiting for a red light when delivering, with the unique goal of volume. But by now, they have found the balance between the volume and the quality, got used to new criterions, and focused on new goals. The number of employees in our store had not increased a lot, but the performance of the whole store had been improved continuously and the quality of service obtained customers' recognition. (Manager of No. 7 Store, Jul 31, 2015)

> At the beginning of the implementation of DD innovation, we could hardly meet the new standards. After a series of specialized training, equipment upgrading and pay system reforming, now we have been used to it. Customers are more satisfied with our service, and in turn our bonuses are also improved. (An employee from No. 7 Store, Jul 31, 2015)

At this time, employees of DD system had been very clear to DD system and the relevant details and standards, and been very familiar with the role that other colleagues played in the work, indeed, formed the recognition of new routines. Meanwhile, the company's new norms and systems provided a clear supporting to participants of the cognition and behavior. With the support of newly-set-up rules and regulations, participants of DD system finally transformed their attention to details and specialty of new tasks, which consequently strengthened new organizational routines with shared schemata for criterions of timing, security and high service quality, and action disposition towards satisfying these criterions. This was a process of new routines solidification, during which new organizational rules and regulations aligned with new routines, and the effect of previously existing routines diminished. Consequently, new routines for new tasks replaced old ones, which indicate successful implementation of DD innovation in HOAU, accompanying with the evolution of routines. Hence fore, HOAU ended the passive situation and opened a new page of development in high-end road transportation market.

5.5 Discussion

5.5.1 The Framework of Routine Evolution in Innovation Implementation

Building on a longitudinal interpretive and exploratory case study of DD innovation in HOAU, and from an implicit perspective of organizational routines, this research confirms management innovation as a system project referring to fundamental change of organizational routines. The research identifies three phases of routine evolution in management innovation implementation, including the existing-organizational-routine-domination phase, the new-routine-creation phase and the new-routine-solidification phase, each exhibiting different innovation activities and participants' cognition and behaviors. Based on the empirical evidence, existing routines dominate cognition and behaviors of participants in the first phase, because the perceived uncertainty stemming from ambiguous innovation goals and proposals makes participants focus on similarity of the new and existing task, follow existing rules and regulations, and hold existing routines. New organizational routines emerge through a process that participants identify the distinction of new tasks, seek to find their new roles through interactions and role taking, and finally form new common understandings and action dispositions. As new rules and regulations are set up and participants move their attention to specialty of new tasks, new organizational routines are solidified. Figure 5.3 graphically displays a framework to illustrate evolution of organizational routines in management innovation implementation.

Accordingly, the author draws the following conclusions:

The research confirms management innovation as a complex project concerning organizational routines which represent the most micro feature of an organization. The success of DD innovation in HOAU relies not only on setting up a good innovation proposal covering a new structure, a new process and a set of new criterions, but importantly on the effective implementation which reflects a process involving all organization levels and especially a fundamental change of the implicit routines. From an explicit perspective, innovation implementation reflects a series of activities taken by the firm, including a preliminary implementation of putting set-up innovation proposals into practices, a deep implementation of improving innovation effectiveness, and a following implementation of establishing new rules and regulations for new tasks; from an implicit perspective, innovation implementation represents fundamental changes of participants' cognition and behaviors in performing new tasks, from focusing on similarity of the new and existing tasks, to their distinction and finally to the specialty of new tasks, from stressing the fitness of existing rules and regulations, to their unfitness, and finally to the fitness of newly established rules and regulations, and also from sticking to existing behavior patterns to creating new patterns, and finally to holding new patterns. These implicit changes demonstrate a replacement of existing organizational routines by new routines, namely, the evolution of organizational routines.

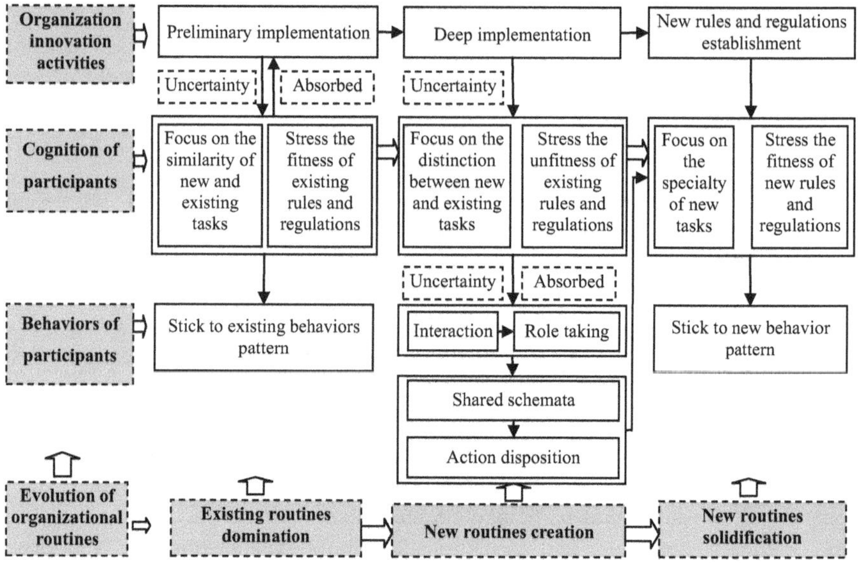

Fig. 5.3 Evolution of organizational routines in management innovation implementation

Explicit innovation activities are downward from the top to the first-line level; while implicit changes occur upward from engaging participants (i.e., the most micro level of an organization) to the top. This indicates that management innovation is a systematical change both downward and upward, rather than an upward strategic issue absolutely relying on top or key managers.

The evolution of organizational routines occurs through a process consisting of existing routines domination, new routines creation and new routines solidification, each exhibiting different innovation activities and characteristics of participants' cognition and behaviors. Despite the stimulation from explicit innovation activities, organizational routines evolve endogenously. The novelty and interruption brought by the systematical innovation at the beginning makes participants perceive a great deal of uncertainty for lack of new information or their inabilities to recognize and interpret relevant information. In order to reduce perceived uncertainty and subsequent anxiety and loss of security, participants in existing routines first strive to turn novelty into familiarity by searching for the similarity between new and the existing context, also to better perform new tasks, at least temporarily. Moreover, since rules and regulations are able to absorb uncertainty, participants tend to stress the fitness of existing rules and regulations for new tasks. As a result, they stick to existing behavior patterns when performing new tasks. This indicates a dominating role of existing routines. With more experience in new tasks, participants adapt to a new context and gradually recognize the actual distinction between new and the existing tasks, and realize the unfitness of existing rules and regulations. Participants even realize the risk of holding existing patterns and begin to adjust their cognition and behaviors. This increases perceived uncertainty of participants, and makes them initiate interactions with others especially those from closely connected links to reconstruct their actions and, in the case of joint action, to fit their actions into the actions of others by assuming their roles or potential responses to different possible actions. Common understandings or shared schemata and action dispositions on new tasks are finally formed through high frequency communications and effective role taking, which indicates the creation of new routines. Participants consolidate their individual schemata and shared schemata by getting rid of pervious habits and skills and forming new ones remained in their procedural memory, with recurrence of performing joint activities. This indicates a dominating role of new routines and a final success of management innovation. All three phases are sequentially linked and proceeds endogenously.

The recreations of new routines are the keys for routine evolution, thus for the success of management innovation. If an innovation stopped at breaking existing routines rather than creating new routines, it would fail to yield the intended results, even make the organization trapped into a new development dilemma. However, creation of new routines for new tasks is a complex issue referring to forming new distinct selves in new joint activities. The unanticipated outcome of management innovation increases routine participants' perceived uncertainty, especially when they realize the unfitness of existing rules and regulations, which results in their going through cycles of interaction to get more information about new tasks. Frequency of interaction facilitates role taking which enables participants to take into account fellow participants' actual and potential understandings, ideas, opinions, and actions with respect to the joint activity, and align their individual lines of action accordingly. This facilitates the development of shared understandings between closely connected participants and finally among all participants. At the same time, context elements or deeper innovation activities help stimulate these internal changes of participants' cognition and behaviors.

5.5.2 *Implications for Theory*

This research focuses on the nature of management innovation implementation and the underlying evolution mechanism of organization routines in implementing innovation, through an exploratory case study on DD innovation of HOAU in China. It makes several important contributions to the literature in both management innovation and organizational routines.

This research breaks the field of organizational innovation to stay in combing the general rules of innovations or refining the obvious impact of the elements of the situation, and extends research on management innovation to most micro level by focusing on innovation implementation. Advocating a positive effect of management innovation on organizational performance and seeing it as one of the most important and sustainable sources of competitive advantage for firms (Hamel 2006; Mol and Birkinshaw 2009), prior studies have explored that why management innovation fails to produce intended results. Most researches focus on examining conditions giving rise to the emergence of management innovation (e.g. Guillén 1994; McCabe 2002; Chi et al. 2007) and identifying sequential phases through which management innovation occurs (e.g. Hamel 2006; Birkinshaw et al. 2008; Lin and Su 2014; Lin et al. 2016). These studies stress upon the importance of previous phases as initiation, decision making and proposal establishment, and focus on effects of various antecedents on these phases, such as institutional factors (Guillén 1994; DiMaggio and Powell 1983), leadership behaviors (Vaccaro et al. 2012), employees (e.g. Kotter and Cohen 2002; Van et al. 2006; Whalen-Berry et al. 2003), the interactions between context and search (Mol and Birkinshaw 2009). This research further reveals the value of implementation phase in innovation and confirms the argument of Birkinshaw et al. (2008) that management innovation potentially reflects fundamental changes in the routines of the organization. More specifically, it confirms implementation, through which the value of the new management practice could be realized, an indispensable phase in management innovation and also a complex project referring to the most micro level of organizational routines. Prior studies gave more attentions to roles of core managers in management innovation (Vaccaro et al. 2012; Lin and Su 2014); while this research shows how employees as participants of routines engage in innovation and make their contributions. It also extends the research of Kotter and Cohen (2002), Van et al. (2006), and Whalen-Berry et al. (2003) who advocate roles of employees, by exhibiting what roles of employees in management innovation are and how they play their roles.

Also, this research contributes to research on the relationship between organizational routines and innovation by taking an organizational routine's perspective to explore management innovation and drawing the conclusion that implementation of management innovation reflects the evolution of organizational routines. Prior researches on innovation have regarded innovation as a systematic project potentially requiring fundamental changes in the routines the organization (e.g. Argyris and Schän 1978; Birkinshaw et al. 2008). At the same time, recent researches on organizational routines have regarded organizational routines as the fundamental to understanding organizational changes or innovation (Becker et al. 2005), or as an important source of flexibility, change and innovation (Nelson and Winter 1982; Pentland and Rueter1994; Feldman 2000; Feldman and Pentland 2003, 2008; Feldman and Pentland 2005; Pentland et al. 2011, 2012). Consistently, this research confirms the relationship between organizational routines and innovation and argues that the success of management innovation relies on evolution of organizational routines. Following Becker et al. (2005), this research advocates the necessary of exploring into the micro-level change in the innovation process. The result indicates that the implicit perspective in management innovation reflects the underlying change of participants' cognition and behaviors, namely the change of shared schemata and action dispositions, or that of organizational routines, a central, micro and fundamental element of organizations (Feldman and Pentland 2003; March and Simon 1958; Cyert and March 1963; Cohen and Klepper 1996; Ortmann 2010). Accordingly, antecedents affect innovation through cognition and behaviors of employees and then evolution of routines. This explains why organizational routines are sources of innovations.

Finally, this research furthers research on organizational routine theory by advocating the two-component view and investigating deeply into the process through which organizational routines evolve. Following Feldman and Pentland's (2003) finding that organizational routines consist of performative aspect and ostensive aspect, and Dionysiou and Tsoukas's (2013) argument that the ostensive aspect is made up of shared schemata and subsequent action dispositions, this research focuses on how existing shared schemata and action dispositions are replaced by new ones endogenously with recurrence of performing new tasks in innovation, namely the evolution of organizational routines. In consist with Dionysiou and Tsoukas (2013), here this research takes a "within" perspective based on cognition and behaviors of participants and stress the roles of interactions and role taking in routine creation and evolution. However, Dionysiou and Tsoukas (2013) do not give specific phases of routine creation; while this research identifies a three-phase process consisting of the existing organizational routines domination, the new routines creation and the new routines solidification. Moreover, Dionysiou and Tsoukas (2013) assume an ideal-typical case of a new organizational setting, with only a few rules or artifacts exerting influence on the initial interaction of participants; while this research is based on a real case of innovation where existing organizational routines need to be overturned and new routines need to be created and solidified, which involves a more complex context. Compared with the four-phase process of formation, inertia, endogenous change, and learning set up by Pentland et al. (2012), or the three-phase process of formation, efficiency, and adaptability by Miller et al. (2012), this process considers not only the changes of participants' cognition and behaviors but also the characteristics of the innovation itself.

5.5.3 Implications for Practice

In addition to contribution to theory, this study offers a number of important insights for innovation practices and relevant managerial practices in firms. Since management innovation is able to bring sustained competitive advantage for a firm, to improve innovation effectiveness especially in implementation is in the interest of managers.

In order to improve innovation, organizations need to enhance participants' personal cognition and collective cognition in order to achieve real innovation. As this research shows management innovation is a systematic project referring to evolution of organizational routines, it is in the interest of firms to call attention of all participants, not only core managers. Besides phases of initiation, decision making and proposal establishment in innovation which rely on top managers to a large extent, innovation implementation is an indispensible phase involving participants even from the first line. Why many firms set up perfect proposals or adopt complete management practices from somewhere else but are still not able to produce intended results may be a lack of attention paid to implementation. Therefore, this research stresses the value of implementation to remind managers that implementation is not a naturally-coming phase. More importantly, besides the explicit changes of the organizational structure, the process or institutions, firms should carefully observe accompanying changes in cognition and behaviors of participants engaged in innovation which reflects the implicit evolution of organizational routines. Only when existing organizational routines are overturned and new routines are created and solidified to further guide participants could we judge that the innovation has been successfully implemented. This reminds managers the complication of innovation implementation.

Though the evolution of organizational routines in innovation is an endogenous process proceeding through interactions and role taking, it is necessary for firms to take actions to appropriately stimulate interactions and role taking among participants, so as to shorten the period of evolution and better realize the goal of innovation. The three phases consist of a complete evolution process of routines. During the previous two phases, management efficiency may be even lower than before and the firms may suffer from a mess, but it's the only path to success in innovation. The existing-organizational-routine-domination phase leads to provide a buffer for organizational innovation or ensure persistence of organization operation and task completion, and the new-routine-creation phase makes an important effect on innovation. This also explains why the enterprise anxious about organizational innovation often ends in failure, there is no time for organizations and participants to gradually get rid of existing routines and form the new ones. Specifically, managers need to make participants better engaged in innovation during the phase of the existing organizational routines domination; to facilitate explicit communications among participants, form role taking and set up shared schemata and action dispositions, during the phase of new routines creation; and to set up new rules and regulations to help solidify new routines in the final phase. That is, managers should understand their changing roles in different stages of routine evolution and take appropriate actions.

5.5.4 Limitations and Future Directions

The effort towards identifying the process through which organizational routines evolve in innovation implementation is constrained by at least three limitations, which also presents opportunities for future research. Being based on a single case study, the findings need to be further confirmed through investigations of other organizations. More cases will help to improve the process framework and further explore the details of organizational routine evolution especially the creation of new routines. Second, implementation of organizational innovation is a complex system problem, this research focuses on the organizational routines' perspective to explore how to successfully implement management innovation; there may be other elements not included, such as cognition of managers, resource orchestration, environmental elements, or organizational culture. It is necessary to carry out further research on the implementation mechanism of organizational innovation from a more comprehensive perspective. Third, when focusing on interactions and role taking in evolution of routines, little attention has been paid to the potential power asymmetries among participants and their potential influence on forming shared schemata and subsequent new routines. Moreover, in developing the process model, this research assumes a small number of participants who interact in conditions of co-presence and only explore the interactions and role taking between or among the links (or groups). Actually, routines may involve a large number of participants, cutting across different levels and functions, and possibly extending beyond the formal boundaries of organizations (Feldman and Pentland 2003; Rerup and Feldman 2011; Turner and Rindova 2012). Such routines are likely to consist of several subroutines and their participants may belong to multiple subgroups. Future research may go deeply into the subroutines to explore how they evolve in innovation or into groups to uncover how group members interact with each other to form common understandings or shared schemata.

5.6 Offers a Conclusion

This chapter addresses how management innovation can be successfully implemented from an organizational routines' perspective. Based on existing literatures on management innovation and organizational routines, this chapter sets up a three-phase framework of organizational routine evolution for innovation by adopting an exploratory case study on DD innovation of HOAU in China. The findings offer many valuable insights for further research in how to improve effectiveness of innovations in special context of China and hold important implications for management practices. Despite its limitations, this research makes several important contributions in emerging literature on management innovation and organizational routines. However, more research should be done for the details of routine evolution and offer more effective paths for improving the performance of management innovation.

References

Anand, G., Gray, J., Siemsen, E.: Decay, shock, and renewal. Organ. Sci. **23**(6), 1700–1716 (2011)

Andriopoulos, C., Lewis, M.W.: Exploitation-exploration tensions and organizational ambidexterity: managing paradoxes of innovation. Organ. Sci. **20**(4), 696–717 (2009)

Argyris, C., Schön, D.A.: Organizational Learning: A Theory of Action Perspective. Birkhäuser Basel (1978)

Ashforth, B.E., Fried, Y.: The mindlessness of organizational behaviors. Hum. Relat. **41**(4), 305–329 (1988)
Balogun, J., Johnson, G.: Organizational restructuring and middle manager sensemaking. Acad. Manag. J. **47**(4), 523–549 (2004)
Barney, J.B.: Firm resource and sustained competitive advantage. J. Manag. **17**(1), 99–120 (1991)
Becker, M.C.: Organizational routines: a review of the literature. Ind. Corp. Change **13**(4), 643–678 (2004)
Becker, M.C., Lazaric, N., Nelson, R.R., Winter, S.G.: Applying organizational routines in understanding organizational change. Post-Print **14**(5), 775–791 (2005)
Birkinshaw, J., Mol, M.J.: How management innovation happens", Sloan Manag. Rev. **47**(4), 81–88 (2006)
Birkinshaw, J., Crainer, S., Mol, M.: Special report on management innovation. Bus. Strategy Rev. **18**(1), 45–73 (2007)
Birkinshaw, J., Hamel, G., Mol, M.J.: Management innovation. Acad Manag Rev **33**(4) (2008)
Birnholtz, J.P., Cohen, M.D., Hoch, S.V.: Organizational character: on the regeneration of camp poplar grove. Organ. Sci. **18**(2), 315–332 (2007)
Blumer, H.: Symbolic Interaction: Perspective and Methods, Englewood Cliffs. Prentice Hall, NJ (1969)
Blumer, H., Morrione, T.J.: George Herbert Mead and Human Conduct. Altamira Press (2004)
Bresman, H.: Changing routines: a process model of vicarious group learning in pharmaceutical R&D. Acad. Manag. J. **56**(1), 35–61 (2013)
Chi, W., Freeman, R.B., Kleiner, M.M.: Adoption and termination of employee involvement programs. NBER working paper, W12878 (2007)
Cohen, M.D., Bacdayan, P.: Organizational routines are stored as procedural memory: evidence from a laboratory study. Organ. Sci. **5**(4), 554–568 (1994)
Cohen, W.M., Klepper, S.: Firm size and the nature of innovation within industries: the case of process and product R&D. Rev. Econ. Stat. **78**(2), 232–243 (1996)
Cronin, M.A., Weingart, L.R.: Representational gaps, information processing, and conflict in functionally diverse teams. Acad. Manag. Rev. **32**(3), 761–773 (2007)
Cyert, R.M., March, J.G.: A behavioral theory of the firm. Prentice-Hall, Englewood Cliffs, NJ (1963)
Damanpour, F., Schneider, M.: Phases of the adoption of innovation in organizations: effects of environment, organization and top managers. Br. J. Manag. **17**(3), 215–236 (2006)
Daneshvar, M., Dikmen, I., Birgonul, M.T.: Investigation of organizational routines in construction companies. Working Paper Proceedings. In: Engineering Project Organizations Conference, Rheden, The Netherlands (2012)
DiMaggio, P.J., Powell, W.W.: The iron cage revisited: institutional isomorphism and collective rationality in organizational fields. Am. Sociol. Rev. **48**(2), 147–160 (1983)
Dionysiou, D.D., Tsoukas, H.: Understanding the (re)creation of routines from within: a symbolic interactions perspective. Acad. Manag. Rev. **38**(2), 181–205 (2013)
Dittrich, K., Guérard, S., Seidl, D.: Talking about routines: the role of reflective talk in routine change. Organ. Sci. **27**(3), 678–697 (2016)
Easterby-Smith, M., Lyles, M.A.: How Should We Understand Tacit Knowledge? A Phenomenological View. Handbook of Organizational Learning and Knowledge Management. Wiley, London (2015)
Eisenhardt, K.M.: Building theories from case study research. Acad. Manag. Rev. **14**(4), 532–550 (1989)
Eisenhardt, K.M.: Better stories and better constructs: the case for rigor and comparative logic. Acad. Manag. Rev. **16**(3), 620–627 (1991)
Eisenhardt, K.M., Graebner, M.E.: Theory building from cases: opportunities and challenges. Acad. Manag. J. **50**(1), 25–32 (2007)
Feldman, M.S.: Organizational routines as a source of continuous change. Organ. Sci. **11**(6), 611–629 (2000)

Feldman, M.S., Orlikowski, W.J.: Theorizing practice and practicing theory. Organ. Sci. **22**(5), 1240–1253 (2011)
Feldman, M.S., Pentland, B.T.: Reconceptualizing organizational routines as a source of flexibility and change. Adm. Sci. Q. **48**(1), 94–118 (2003)
Feldman, M.S., Pentland, B.T.: Organizational routines as a unit of analysis. Ind. Corp. Change **14** (5), 793–815 (2005)
Feldman, M.S., Pentland, B.T.: Designing routines: on the folly of designing artifacts, while hoping for patterns of action. Inf. Organ. **18**(4), 235–250 (2008)
Feldman, M.S., Pentland, B.T., D'Adderio, L., Lazaric, N.: Beyond routines as things: introduction to the special issue on routines dynamics. Post-Print **27**(3), 505–513 (2016)
Felin, T., Foss, N.J.: Organizational routines a sceptical look. Druid Working Papers, 04–13 (2004)
Geiger, D., Schröder, A.: Ever-changing routines? Toward a revised understanding of organizational routines between rule-following and rule-breaking. Schmalenbach Bus. Rev. **66**(2), 170–190 (2014)
Gersick, C.J., Hackman, J.R.: Habitual routines in task-performing groups. Organ. Behav. Hum. Decis. Process. **47**(1), 65–97 (1990)
Glaser, B.G., Strauss, A.L.: The Discovery of Grounded Theory: Strategies for Qualitative Research. Aldine, Chicago (1967)
Golden-Biddle, K., Locke, K.: Composing Qualitative Research: Crafting Theoretical Points from Qualitative Research. Sage, CA (2007)
Guillén, M.F.: Models of Management: Work, Authority, and Organization in A Comparative Perspective. University of Chicago Press, Chicago (1994)
Gummesson, E.: Qualitative Research Methods in Management Research. Sage, Thousand Oaks, CA (2000)
Hage, J.: Theories of Organization: Form, Process, and Transformation. Wiley, New York (1980)
Hamel, G.: The why, what, and how of management innovation. Harv. Bus. Rev. **84**(2), 72–84 (2006)
Hamel, G.: The Future of Management. Harvard Business School Press, Boston (2007)
Hannan, M.T., Freeman, J.R.: Structural inertia and organizational change. Am. Sociol. Rev. **49** (2), 149–164 (1984)
Heimeriks, K.H., Schijven, M., Gates, S.: Manifestations of higher-order routines: the underlying mechanisms of deliberate learning in the context of postacquisition integration. Acad. Manag. J. **55**(3), 703–726 (2012)
Hodgson, G.M.: In: Becker, M.C. (ed.) The Concept of A Routine. Handbook of Organizational Routines. Edward Elgar Publishing, pp. 14–30 (2008)
Howard-Grenville, J., Rerup, C., Langley, A., Tsoukas, H.: Organizational Routines: How they Are Created, Maintained, and Changed. Oxford University Press (2016)
Howard-Grenville, J.A.: The persistence of flexible organizational routines: the role of agency and organizational context. Organ. Sci. **16**(6), 618–636 (2005)
Joas, H.: GH Mead: A Contemporary Re-examination of His Thought. MIT Press (1997)
Kotter, J.P., Cohen, D.S.: Creative ways to empower action to change the organization: cases in point. J. Organiz. Excell. **22**(1), 73–82 (2002)
Krippendorff, K.: Content Analysis: An Introduction to Its Methodology, 2nd edn. Sage, Thousand Oaks. CA (2004a)
Krippendorff, K.: Reliability in content analysis. Hum. Commun. Res. **30**(3), 411–433 (2004b)
Langley, A.: Strategies for theorizing from process data. Acad. Manag. Rev. **24**(4), 691–710 (1999)
Lazaric, N.: In: Becker, M.C. (ed.) Routines and Routinization: An Exploration of Some Micro-Cognitive Foundations. Handbook of Organizational Routines. Edward Elgar Publishing, pp. 225–227 (2008)
Lin, H.F., Su, J.Q.: Research on the introducing mechanism of management innovations based on the perspective of internal promoters. Sci. Res. Manag. **31**(6), 136–143 (2010)

Lin, H.F., Su, J.Q.: A case study on adoptive management innovation in China. J. Organiz. Change Manag. **27**(1), 83–114 (2014)
Lin, H.F., Su, J.Q., Higgins, A.: How dynamic capabilities affect adoption of management innovations. J. Bus. Res. **69**(2), 862–876 (2016)
Lincoln, Y.S., Guba, E.G.: Naturalistic Inquiry Sage. Sage Publications, Beverly Hills, CA (1985)
Loch, C.H., Sengupta, K., Ahmad, M.G.: The microevolution of routines: how problem solving and social preferences interact. Organ. Sci. **24**(1), 99–115 (2013)
March, J.G.: Exploration and exploitation in organizational Learning. Organ. Sci. **2**(1), 71–87 (1991)
March, J.G., Simon, H.A.: Organizations. Wiley, New York (1958)
Marks, M.L.: Workplace recovery after mergers, acquisitions, and downsizings: facilitating individual adaptation to major organizational transitions. Org. Dyn. **35**(4), 384–399 (2006)
Mccabe, D.: 'Waiting for Dead Men's Shoes': towards a cultural understanding of management innovation. Hum. Relat. **55**(5), 505–536 (2002)
Mead, G.H.: Mind, Self, and Society from the Standpoint of a Social Behaviorist. University of Chicago, Chicago (1934)
Meaney, M., Pung, C.: McKinsey global results: creating organizational transformations. McKinsey Q. **7**(3), 1–7 (2008)
Miller, D.L.: George Herbert Mead: self, language, and the world. American J. Sociol. (1973)
Miller, K.D., Pentland, B.T., Choi, S.: Dynamics of performing and remembering organizational routines. J. Manage. Stud. **49**(8), 1536–1558 (2012)
Miner, A.S.: Structural evolution through idiosyncratic jobs: the potential for unplanned learning. Organ. Sci. **1**(2), 195–210 (1990)
Mol, M.J., Birkinshaw, J.: The sources of management innovation: when firms introduce new management practices. J. Bus. Res. **62**(12), 1269–1280 (2009)
Narduzzo, A., Rocco, E., Warglien, M.: Talking about routines in the field: the emergence of organizational capabilities in a new cellular phone network company. Ceel Working Papers, pp. 27–51 (1997)
Nelson, R.R., Winter, S.G.: An Evolutionary Theory of Economic Change. Belknap Press/Harvard University Press, Cambridge (1982)
Orlikowski, W.J.: Knowing in practice: enacting a collective capability in distributed organizing. Organ. Sci. **13**(3), 249–273 (2002)
Ortmann, G.: On drifting rules and standards. Scand. J. Manag. **26**(2), 204–214 (2010)
Paper, D., Chang, R.D.: The state of business process reengineering: a search for success factors. Total Qual. Manag. Bus. Excell. **16**(1), 121–133 (2005)
Parmigiani, A., Howard-Grenville, J.: Routines revisited: exploring the capabilities and practice perspectives. Acad. Manag. Ann. **5**(1), 413–453 (2011)
Pentland, B.T., Feldman, M.S., Becker, M.C., Liu, P.: Dynamics of organizational routines: a generative model. Social Sci. Electron. Publ. **49**(8), 1484–1508 (2012)
Pentland, B.T., Hærem, T., Hillison, D.: Comparing organizational routines as recurrent patterns of action. Organiz. Stud. **31**(7), 917–940 (2010)
Pentland, B.T., Rem, T., Hillison, D.: The (N)ever-changing world: stability and change in organizational routines. Organ. Sci. **22**(6), 1369–1383 (2011)
Pentland, B.T., Rueter, H.H.: Organizational routines as grammars of action. Adm. Sci. Q. **39**(3), 484–510 (1994)
Pierce, J.L., Delbecq, A.L.: Organization structure, individual attitudes and innovation. Acad. Manag. Rev. **2**(1), 27–37 (1977)
Pil, F.K., Macduffie, J.P.: The adoption of high-involvement work practices. Ind. Relat. J. Econ. Soc. **35**(3), 423–455 (1996)
Quinn, R.E.: Building The Bridge as You Walk on It: A Guide for Leading Change. Leader to Leader. Jossey-Bass Publishers (2004)
Rentsch, J.R., Small, E.E., Hanges, P.J.: Cognitions in organizations and teams: what is the meaning of cognitive similarity? In: Smith, D.B. (ed.) The People Make the Place: Dynamic

Linkages Between Individuals and Organizations, pp. 127–155. Lawrence Erlbaum Associates, New York (2008)
Rerup, C., Feldman, M.S.: Routines as a source of change in organizational schemata: the role of trial-and-error learning. Acad. Manag. J. **54**(3), 577–610 (2011)
Rogers, E.M.: Diffusion of Innovations, 4th edn. The Free Press, New York (1995)
Salvato, C., Rerup, C.: Beyond collective entities: multilevel research on organizational routines and capabilities. J. Manag. **37**(2), 468–490 (2011)
Stene, E.O.: An approach to a science of administration. Am. Polit. Sci. Rev. **34**(6), 1124–1137 (1940)
Strauss, A., Corbin, J.M.: Basics of Qualitative Research: Grounded Theory Procedures and Techniques. Sage, Newbury Park, CA (1990)
Teece, D.J., Pisano, G., Shuen, A.: Dynamic capabilities and strategic management. Strateg. Manag. J. **18**(7), 509–533 (1997)
Thompson, T.E.: Direct development in a nudibranch, cadlina laevis, with a discussion of developmental processes in opisthobranchia. J. Mar. Biol. Assoc. U.K. **47**(1), 1–22 (1967)
Tsoukas, H., Chia, R.: On organizational becoming: rethinking organizational change. Organ. Sci. **13**(5), 567–582 (2002)
Tucker, A.L., Edmondson, A.C.: Why hospitals don't learn from failures: organizational and psychological dynamics that inhibit system change. Calif. Manag. Rev. **45**(2), 55–72 (2003)
Turner, S.F., Rindova, V.: A balancing act: how organizations pursue consistency in routine functioning in the face of ongoing change. Organ. Sci. **23**(1), 24–46 (2012)
Vaccaro, I.G., Jansen, J.J.P., Bosch, F.A.J.V.D., Volberda, H.W.: Management innovation and leadership: the moderating role of organizational size. J. Manage. Stud. **49**(1), 28–51 (2012)
Van, K.B., Martin, L., Tyler, T.: Process-orientation versus outcome-orientation during organizational change: the role of organizational identification. J. Organiz. Behav. **27**(6), 685–704 (2006)
Weiss, H.M., Ilgen, D.R.: Routinized behavior in organizations. J. Behav. Econ. **14**(1), 57–67 (1985)
Wernerfelt, B.: A resource-based view of the firm. Strateg. Manag. J. **5**(2), 171–180 (1984)
Whalen-Berry, K.S., Gordon, J.R., Hinings, C.B.: Strengthening organizational change processes: recommendations and implications from a multi-level analysis. J. Appl. Behav. Sci. **39**(2), 186–207 (2003)
Winter, S.G.: Toward a neo-schumpeterian theory of the firm. Soc. Sci. Electron. Publ. **15**(1), 125–141 (2006)
Yin, R.K.: Case Study Research: Design and Methods, 2nd edn. Sage, Newbury Park, CA (1994)

Chapter 6
How Do Dynamic Capabilities Drive Adoptive Management Innovation?

6.1 Introduction

Following the process framework set up in Chap. 3, and studies of Vaccaro et al. (2012), Mol and Birkinshaw (2009) and Lin and Su (2014), this chapter furthers the exploration on the driver of adoptive management innovation from the perspective of dynamic capability. Besides institutional factors (Guillén 1994; DiMaggio and Powell 1983), leadership behaviors (Vaccaro et al. 2012), and the interaction between context and search (Mol and Birkinshaw 2009), management innovation also relies on the whole organizational system with valuable, rare, inimitable, and non-substitutable resources (Barney 1991; Teece et al. 1997; Wernerfelt 1984). Accumulating valuable resources is insufficient to support sustainable competitive advantages in the ever-changing competitive environment (Teece et al. 1997, Teece 2007; Liao et al. 2009). Dynamic capabilities are capabilities of integrating, reconfiguring, gaining and releasing these resources (Teece et al. 1997; Eisenhardt and Martin 2000; Zollo and Winter 2002) which may enable firms to reconfigure internal and external competencies to support management addressing the challenges faced in rapidly changing environments. Succinctly, management innovation alone is insufficient to generate success (Teece 1986, 2007) without the dynamic capabilities of a firm to purposefully create, extend or modify its resource base.

In practice, although many companies are trying to manage a variety of management innovation, the results are unsatisfactory. It happens very often that implementing business process reengineering and diversification strategy leads to the whole organization not only fail to achieve the transformation but also be immersed in an embarrassing situation. Therefore, the systematic promotion of efficiency of management innovation is an urgent problem to be solved in the field of current management innovation. The effectiveness of management innovation is influenced by multidimensional factors, which means that the path of promotion is also diversified. Actually, management innovation is closely related to the whole organization's operating system, and it is impossible to achieve it overnight since it

H. Lin, *Adoptive Management Innovation*,
https://doi.org/10.1007/978-981-10-7712-8_6

is a continuous system, which is the fundamental guarantee of business resources and capacity resources. Dynamic capability is a complex concept, but from the nature of dynamic capability, it emphasizes the enterprise through the integration, use, recycling resources to create new competitive capacities to achieve the purpose of matching with the external environment. Innovation is an important way to realize dynamic dominance. Particularly, management innovation with high demand for resources and information is undoubtedly an important weapon for organizations to cope with environmental change, solve internal efficiency problems and maintain competitiveness to show the dynamic capability of an organization. In turn, the improvement of organizational dynamic capability may also promote the improvement of the effectiveness of organizational management innovation, but the complexity of dynamic capabilities determines its management innovation effectiveness of the mechanism, and further enhances the complexity of the mechanism.

Seeing dynamic capabilities as an internal driver for management innovation and focusing on what makes adoptive management innovation distinctive, this chapter attempts to address two questions: first, how can adoptive management innovation be measured as a complex and multidimensional concept or process where results are intangible, uncertain, lagging and even inseparable from that of technological innovation? Considering this difficulty in measuring outputs of management innovation, this research advocates a process-oriented method based on the set-up of a four-phase framework instead of a results-oriented one based on ambiguous outputs. Therefore, this section sets up a measuring scale based on this process-oriented method, by extracting characteristics and major activities in each phase of innovation. Second, how can dynamic capabilities efficiently enhance the process of adoptive management innovation? Previous research emphasizes the relationship between dynamic capabilities and innovation (e.g. Teece et al. 1997, 2007; Helfat et al. 2007; Ridder 2011; Aragón-Correa and Sharma 2003; Hart and Dowell 2010; Clausen 2013; Ambrosini et al. 2009; Kohlbacher 2013; Cheng and Chen 2013), and contends a firm's dynamic capabilities could significantly enhance its ability to innovate (O'Connor 2008), especially in the case of radical management innovation. Nonetheless, the literature still contains gaps in discussing how dynamic capabilities internally enhance performance of management innovation. Closer analysis also reveals the majority of research is theoretical and conceptual. Therefore, this section attempts to explore how different dynamic capabilities affect each phase of adoptive management innovation by using Partial Least Squares Structural Equation Modeling (PLS-SEM), to address the lack of empirical evidence and offer managerial implications.

This chapter is organized as follows. Section 6.2 describes the measuring and process of management innovation. Section 6.3 describes a literature review on effects of dynamic capability, and components of dynamic capability. Section 6.4 presents hypotheses and the theoretical framework (also inner model) reflecting relationships between dynamic capability and four phases of adoptive management innovation. Section 6.5 presents research setting, data collection, measures and the measurable model (also outer model). Section 6.6 shows the findings obtained from

Partial Least Squares Structural Equation Modeling (PLS-SEM) based on 264 respondents from China. Section 6.7 presents a discussion of the findings and contributions, and managerial implications, and Sect. 6.8 offers a brief conclusion.

6.2 Measurement of Adoptive Management Innovation

Management innovation has gradually become the most important and continuous source of competitive advantages, but its complexity and lag of impact on organizational performance make the scholars and industrialists puzzled on how to improve the effectiveness of management innovation. Though both the significant effects of management innovation on organizational performance and their high research value have been identified (Armbruster et al. 2008), the measurement is lacking. Mol and Birkinshaw (2009) and Vaccaro et al. (2012) use a results-oriented method from the tangible technological innovation field (e.g. Souitaris 2001; Romijn 2002) to measure management innovation. However, Birkinshaw et al. (2008, p. 829) argue that "there are important differences in the nature of the outputs of management innovation and technological innovation". For example, unlike technological innovation, management innovation is typically tacit in nature. They are also relatively difficult to observe, identify system borders, whilst results are lagging and even inseparable (Teece et al. 1997). Moreover, as Birkinshaw et al. (2008) further argue, these attributes increase the importance of the process. Such attributes make it difficult to measure the results of management innovation quantitatively, and Mol and Birkinshaw (2009) suggest more research focus on these poorly understood processes. In fact, both the process of generating management innovation set up by Birkinshaw et al. (2007, 2008) and that of adoptive management innovation by Lin and Su (2014) indicate the importance of innovation process in realizing its goal of improving organizational performance. As for the measurement of management innovation effectiveness or innovation capability, researches have mainly focused on technological innovation, and even though some studies have paid close attention to organizational innovation behavior in general, it still emphasizes the dominant position of technological innovation. Part of researches argue that the performance perspective measurement method cannot fully consider a large number of hidden factors of technical innovation. In the measurement of technological innovation effectiveness, performance perspective and process perspective are both reasonable. But in terms of management innovation, the results have ambiguousness, hysteresis, non-separability and other characteristics, so the process perspective is more suitable for management innovation effectiveness measure. Therefore, different from the result-oriented method proposed in Chap. 4, here the author adopts a process-oriented method for measuring which would be more suitable for exploring deeply into the nature of management innovation.

Then, what would be the process through which adoptive management innovation occurs? The set up framework in the Chap. 3 highlights the roles of key staff and even their mental activities; however, considering the reliance on the whole organization for successful initiation to implementation of management innovation, the author takes the organizational perspective in this chapter. Therefore, this research redevelops a framework of adoptive management innovation based on previous research and the case study of the market-chain-based Business Process Reengineering of Haier Group in China[1] because "a considerable proportion of management research is based on casework, inducting generalizations from specific instances of documented managerial practice" (Tranfield and Starkey 1998). The use of case studies for theorizing, modeling and research is referred to as 'Mode 2' research where there is a "constant flow back and forth between the fundamental and the applied, between the theoretical and the practical" (Gibbons et al. 1994, p. 19).

According to Lin (2009), Haier began to implement the market-chain-based business process reengineering system in late 1998. A market chain is a series of business process activities to make products or render services to satisfy customers' needs. In a nutshell, a market chain links every employee's work with the market, which can be an external or internal market. Therefore, every Haier employee's next downstream activity or process is a market, and every employee faces a market with a direct link to a customer. This allows the firm to convert external market competition into a type of internal competition. Therefore, with employee compensation tied to market performance, every employee provides the best performance to meet his or her customers' needs. To do so, every Haier employee has a picture of the entire organization that shows how company parts interrelate. For example, the production department's direct customer is the distribution department. If you ask an employee where an order comes from, he or she can tell you. To understand the company's entire market-chain system, each employee attends training at Haier University and learns everything from product development to production and distribution.

Figure 6.1 shows the synchronous flow model of Haier's market chains. The top row shows the management process of strategic planning, operation reporting, internal audit, and process and IT management. The second row shows the supply chain planning that links with both supplier relationship management (SRM) to obtain the best global supply chain resource and customer relationship management (CRM) to provide excellent service to global customers. There are three major

---

[1]Case sources: Chinese Management Awards (2006) by PKU Business Review, http://www.haier.net/cn, and Lin 2009, 91:41–49. Haier Global (Haier) was founded in 1984 in Qingdao, in China. Through its entrepreneurial and innovative spirit, Haier has transformed itself from an insolvent collectively-owned factory on the brink of bankruptcy into the number one global home appliance brand in the last 30 years. Haier is known for disruptive innovation in its product solutions and management model, e.g. the OEC management control system, unique performance management systems, the market-chain-based business process reengineering system and Win-win Model of Individual-Goal Combination. In particular, Haier's restructuring effort on the market-chain-based business process reengineering system which began in late 1998 has taken it from a nearly bankrupt factory to a company with global revenue of RMB180.3 billion (USD 29.5 billion) in 2013.

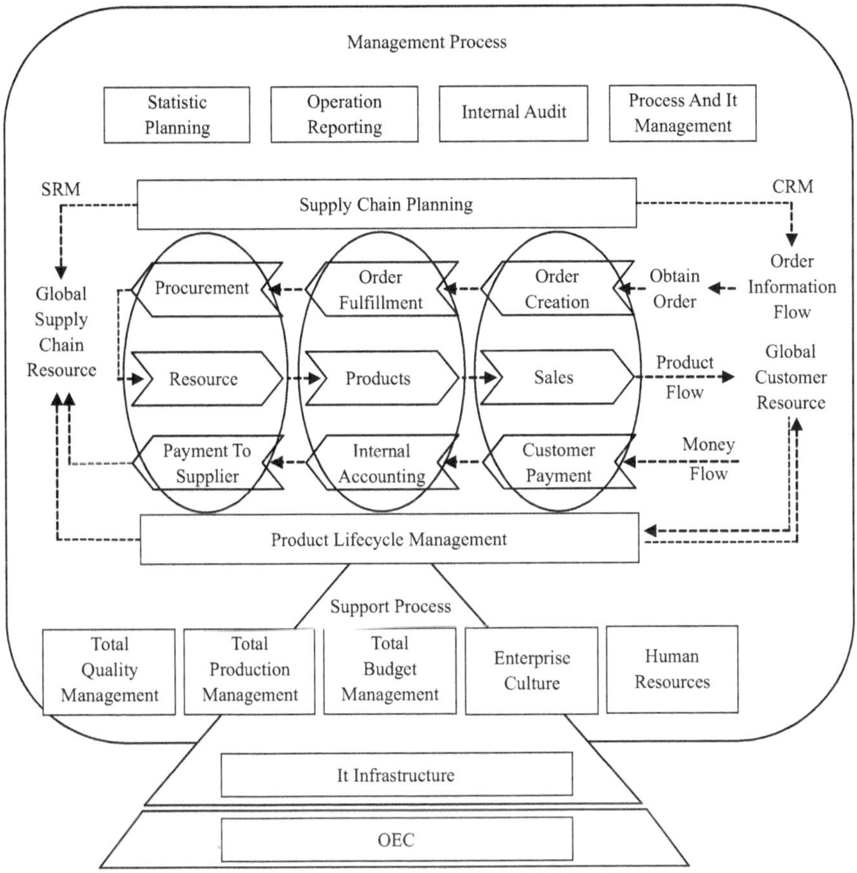

Fig. 6.1 Market-chain-based business process reengineering system of Haier. *Data source* Lin (2009), 41–49

flows: order information, product, and money flow. In the center of the diagram, there are three circles. The left circle shows the primary activities of the logistics division. The middle circle shows the primary activities of the various product divisions. The right circle shows the primary activities of the marketing and sales division. The company pays attention to product lifecycle management (PLM). All service departments support the three circles with total quality management, total production management, total budget management, enterprise culture, and human resources management. The fundamental bases of the reengineering system are the IT infrastructure and Haier's OEC management-control system.

Through the case study on Market-Chain-Based Business Process Reengineering System of Haicr, a four-phase process consisting of initiation, outside search, proposal establishment, and innovation implementation is reset up in this chapter.

Initiation reflects a firm's recognition of the need for management innovation: sensing a novel problem and initial judgment. Through identifying and analyzing external environmental changes, they impact the business and the problems in the organization's internal operations to find opportunities to develop innovative activities in organizations. Guillén (1994) and Birkinshaw et al. (2008) suggest the demand for new management practices is driven by novel problems leading to low efficiency. For example, Haier initiated a change in its systems when sensing e-commerce challenges and China's accession to the World Trade Organization and dealing with these new relationships. Such initiations lead to an outside search. Outside search refers to assimilating the management innovation experiences and practices of other enterprises from different innovative sources of knowledge.

In this phase, a problem-driven search obtains an existing management practice (Williams and Rao 1998), recognizes an opportunity to integrate it (Wilson 1987; Zaltman et al. 1973), and seeks further information from their networks to absorb it. Haier's search focused on existing philosophies of the efficient market-chain system and order-process performance. The firm then explores problem sources and establishes new processes: the phase of proposal establishment. Proposal establishment means proposing a new program for the organization's own conditions., that is, the basis for the absorption of external knowledge to propose new management innovation measures in order to match the problems with the organization.

Integrating these adopted philosophies with existing problems, Haier set up an reengineering system consisting of two flows, as shown in Fig. 6.1. Proposal establishment indicates that adoption of management practices is not simply knowledge transfer but a more complicated and logical process of integration within this new organizational context.

At the implementation phase, organizations need to adjust and integrate organizational resources, coordinate efforts to jointly practice management innovation. Haier first spent six months assisting staff to absorb these innovations before transforming the Group's pyramid structure into a matrix structure focused on project operations; then, creating three major interactive processes with divisions under each. Under the old, only the sales departments had to face the market directly. Before the change, Haier's organizational structure, a pyramid structure, was as follows: The headquarters was the planning center; The product-line divisions were investment centers or profit centers; The sales departments were revenue centers; The factories and service departments were cost; and The work teams were the quality centers. In March 1999, Haier began to transform the Group's pyramid structure into a matrix structure. Under this matrix, the horizontal axis consists of functional departments, and the vertical axis consists of projects. The new structure maintained all the divisions and their R&D, procurement, and sales departments, but the divisions now needed to interact with other divisions on certain projects. From mid-August to October 1999, Haier implemented a revolutionary organizational change which created three major interactive processes with divisions under each: development or core, functional or supporting, and product. As Birkinshaw et al. (2008) and Lin and Su (2014) suggest, this phase of innovation implementation also included activities on the "technical" side to realize the value of management innovation by putting the proposal into practice.

Similarly, in the case of the successful implementation of Jiangxi Mobile to introduce the process of management innovation, the author found out that the successful implementation of its management innovation was also can be divided into four steps. First of all, Jiangxi Mobile managers focused on the management of the opportunity to identify the initial identification and found that Jiangxi Mobile was both facing internal development needs and external environmental pressures two major challenges. On the one hand, as for internal development needs, with the continuous expansion of the scale of enterprises, "the big business disease" gradually brewed, for example institutions bloated, serious barriers, slow information transmission, coordination ineffective, bureaucratic appearance, entrepreneurial passion and other symptoms, and all of these might become the obstacles in future development of enterprises. On the other hand, when it comes to the pressure from the external environment, with the application of 3G and the reorganization of telecom enterprises, the development of enterprises was faced with many uncertain factors such as triple play, value chain fusion, business and content integration, and the homogeneity of communication market in China was becoming more and more obvious, the management level would become the main source of competitive advantage, then they decided to initiate innovation; Secondly, in view of the existing problems or challenges of the organization, managers examined the existing innovative practices or methods already found, and found that the efficiency management as a basic goal of management under the strategic goal might be an effective way to solve the existing problems of the company, so they made the decision-making of the introduction of efficiency management. Then, Jiangxi Mobile, driven by efficiency management as a strategic goal, through the construction of systematic diagnosis, concluded that there were different problems in the process management(part of the important process was missing or outdated, or no curing; part of the process did not form a customer-oriented process closed-loop management; part of the process chain were too long, and resulted in the failure to respond to competitors in the challenges and opponents), performance management, quality management, risk management, organization management and IT support system, and formulated innovation plan according to the efficiency management idea and method. On the one hand, they constructed the framework of efficiency management theory system through process management, organization management, performance management, knowledge management, quality management and risk management. On the other hand, they formulated specific promotion measures, including process promotion, that is, the establishment of a customer-centric process system, the optimization of some key processes, through the vertical and horizontal optimization processes to inspect process efficiency, and the establishment of long-term mechanism of process management. At the same time organizational improvement, which was to break the organization of horizontal barriers, improved the vertical organization, to promote the organization's and the company's development match; performance improvement, which was to improve the KPI settings, promoted cross-sector cooperation, improved staff careers, trainings and evaluation mechanisms. Finally, according to the innovative program, the full mobilization of organizational resources, Jiangxi Mobile spent nearly three years to implement efficiency management.

This four-phase framework of adoptive management innovation offers a base for setting up a process-oriented measuring scale in the following. Nevertheless, it should be noted that, although implementation often represents where outputs of management innovation come from, this chapter focuses on the process of implementation (i.e., how innovations are implemented) rather than outputs (i.e., results produced by implementation or outputs of management innovation). Therefore, this research selected indicators reflecting the efficiency of putting innovation into practice for measuring the performance of implementation, for example, the support from the whole organization, and the predicted progress.

6.3 Dynamic Capabilities and Management Innovation

6.3.1 The Relationship Between Dynamic Capabilities and Management Innovation

Though abundant studies have been conducted on management innovation, the problem of failing to yield intended results still has not been addressed. Management innovation is a vast, complex and enduring high risk project which relies on a whole organizational system with valuable, rare, inimitable, and non-substitutable resources (Barney 1991; Teece et al. 1997; Wernerfelt 1984). Nonetheless, accumulating these valuable resources is insufficient to support a sustainable process of management innovation in a dynamic competitive environment (Teece et al. 1997, 2007; and Liao et al. 2009). Teece (2007) specifically suggests that the sustainable innovation requires not only the ownership of these difficult-to-replicate resources but also unique and difficult-to-replicate dynamic capabilities which are defined as "the firm's ability to integrate, build and reconfigure internal and external competencies to address rapidly changing environments" (Teece et al. 1997, p. 515). Further, gaining and releasing resources (Eisenhardt and Martin 2000) to renew management processes enhances operational performance and integration of new environmental requirements (e.g. Teece et al. 1997, 2007; Eisenhardt and Martin 2000; Helfat and Peteraf 2003; Helfat et al. 2007; Ridder 2011). In other words, the dynamic capabilities by virtue of a firm's people or material resources (Teece et al. 1997; Eisenhardt and Martin 2000; Zollo and Winter 2002), are essential in implementing innovations in an effective and efficient process. Accordingly, dynamic capabilities are able to facilitate the process of initiating and implementing feasible management innovation. Specifically, management innovation alone is insufficient for generating success (Teece 1986) without the dynamic capabilities of a firm to purposefully create, extend or modify its resource base.

When highly dynamic and unpredictable environments make a firms' existing competences quickly obsolete, dynamic capabilities need to be implemented to rebuild competitive resource bases and innovative management systems in a timely and astute manner (Eisenhardt and Martin 2000; Helfat et al. 2007). Moreover, management innovation is needed to respond to fundamental changes in the competitive environment or to reposition the firm (Brady and Davies 2004). An infinite spiral of dynamic capabilities to renew existing resources and capabilities is needed for this innovation with a change in how the firm solves its problems. Given that the context in which firms deal with various emerging sustainable issues is highly complex and sometimes ambiguous, some studies suggest dynamic capabilities can be applied throughout the entire process by which firms undertake sustainable innovation (e.g. Aragón-Correa and Sharma 2003; Hart and Dowell 2010). For example, Clausen's (2013) research showed both direct and indirect relationships between dynamic capabilities and management innovation; Kohlbacher (2013) empirically examined the impact of dynamic capabilities on innovation performance through continuous improvement; Lazonick and Prencipe (2005) sought to analyze the role of dynamic capabilities in sustaining the entire innovation process, and argued that innovation depends on "strategic control" and "financial commitment" (i.e., two dynamic capabilities); Liao et al. (2009) argue the firm's ability to mobilize its resources and capabilities and align them dynamically with changing opportunities is of vital importance to constantly innovate, survive and create a competitive advantage; Wu et al. (2012) showed that dynamic capabilities facilitated firms' strategic changes towards sustainability and higher competitive advantage through searching, prioritizing, positioning, planning, modifying, and leveraging. In summary, researchers have emphasized the relationship between dynamic capabilities and management innovation throughout the process, and a firm's dynamic capabilities could significantly enhance its ability to innovate, especially in the case of the radical management innovation (O'Connor 2008).

The primary premise is that a firm has operational capabilities and resources that are directly involved in the innovation process by converting inputs into outputs, and dynamic capabilities influence this process by updating, integrating and reconfiguring a firm's existing operational capabilities and resources (Helfat et al. 2007; Helfat and Peteraf 2003; Helfat and Winter 2011). In contrast to dynamic capabilities, operational capabilities and resources are static ("zero-order") abilities in the sense that they cannot change unless they are acted upon by dynamic capabilities (Collis 1994; Helfat and Winter 2011; Winter 2003). As such, the effects of dynamic capabilities on the innovation process have progressed into two research streams. The first stream investigates how firms use dynamic capabilities to reconfigure, build and integrate zero-order operational capabilities that support innovative activities (e.g. Helfat 1997; Kogut and Zander 1992; Teece et al. 1997; Winter 2003; Zollo and Winter 2002). The other stream investigates how firms use dynamic capabilities to reconfigure tangible and intangible resources throughout the innovation process, especially complex, large-scale management innovation projects (Eisenhardt and Martin 2000; Helfat and Peteraf 2003). This research contributes to the latter stream.

Though the positive effects of dynamic capabilities on innovation have been extensively identified, the literature still contains research gaps in how dynamic capabilities specifically enhance the internal performance of management innovation. Further, a closer look at the literature reveals that the majority are theoretical and conceptual. Therefore, this study attempts to explore how different dynamic capabilities affect each phase of the adoptive management innovation, thereby addressing this research gap, and answering the second research question: how could dynamic capabilities efficiently enhance performance of the adoptive management process?

6.3.2 Components of Dynamic Capabilities

Some studies measure dynamic capabilities as a single-dimensional construct, for example, integrative capabilities (Liao et al. 2009) or absorptive capacities (Cheng and Chen 2013). However, the majority of studies advocate a multidimensional construct. Teece et al. (1997, p. 510) make "rudimentary efforts…to explain how combinations of competences and resources can be developed, deployed" but define capabilities as adapting, integrating, reconfiguring skills and resources assembled in integrated clusters, and their researches serve as a basis for more detailed modeling of dynamic capabilities. As Zahra and George (2002) suggest, firms have different dynamic capabilities that serve different purposes and functions. Similarly, the complicated process of adoptive management innovation needs support from various dynamic capabilities, so the study here advocates the multidimensional construct view.

Then, what are the components of a firm's dynamic capability? Different studies emphasize different purposes. Teece (2007) argues that the dynamic capability of a firm could be disaggregated into: the capacity to sense and shape opportunities and threats (i.e., sensing capability); to seize opportunities (i.e., seizing capability); and to maintain competitiveness through enhancing, combining, protecting, and reconfiguring the business enterprise's intangible and tangible assets (i.e., reconfiguring capability). Consistent with the process of a firm's strategic management from scanning through to implementation, to directly serve strategic creation and management, the research of Teece (2007) enables managers to delineate strategic considerations and the priorities they must adopt to enhance enterprise performance, and escape the zero-profit tendency. Therefore, such research reflects a macro-level of analysis emphasizing the key role of dynamic capabilities in strategic management (Teece et al. 1997). From a similar macro level, and seeing dynamic capabilities as the antecedent organizational routines by which managers alter their resource deployment to generate new value-creation strategies (Eisenhardt and Martin 2000), Wu et al. (2012) identify three distinct but related dynamic capabilities: scanning, identification, and reconfiguration capabilities. From a micro-level of analysis, Helfat and Peteraf 2003, Helfat et al. (2007) argue that dynamic capabilities could be people-resource based or material-resource based. Analyzing 13 empirical studies on dynamic capabilities, Wang and Ahmed (2007) identify three components of adaptive, absorptive and innovative capabilities to explain the link between internal resource advantage and external marketplace-based competitive advantage. Since management innovation may rely on both macro- and micro-level of dynamic capabilities (for example, sensing both external changes and internal problems, establishing both internal and external relationships), this study identifies components based on dimensions of internal capabilities vs. external capabilities, and dimensions of sensemaking versus sensegiving (Gioia and Chittipeddi 1991). A Fuzzy Cluster Analysis of 62 dynamic capabilities (or capability) articles published in international academic journals (e.g. Helfat 1997, Helfat et al. 2007; Kogut and Zander 1992; Teece 2007; Teece et al. 1997; Wang and Ahmed 2007; Winter 2003; Wu et al. 2012; Zollo and Winter 2002) is conducted to identify the components. Results indicate four distinct but interacting components: sensing capability for directional change (e.g. Teece 2007; Wang and Ahmed 2007; Wu et al. 2012), absorptive capacity for organizational learning (e.g. Cheng and Chen 2013; Davenport et al. 2006; Hertog et al. 2010; Wang and Ahmed 2007), relational capability for relationships and social capital acquisition (e.g. Barreto 2010; Helfat et al. 2007), and integrative capability for communication and coordination (e.g. Liao et al. 2009; Kogut and Zander 1992; Teece 2007; Winter 2003; Wu et al. 2012; Zahra et al. 2006), as appearing in Fig. 6.2.

Each component has a particular emphasis: the sensing capability is a cognitive or mental process through which organizations perceive environmental changes and how to implement them (for example, opportunity identification capability, and adaption capability); the absorptive capacity is the ability to acquire knowledge and transform the acquired knowledge into firm-embedded knowledge through assimilation, transformation and exploitation capability; the relational capability is a firm's ability to build relationships and acquire resources from the relationships (for example, social capital acquisition, social-relationship building, and interaction-promoting capability); and the integrative capability refers to the ability

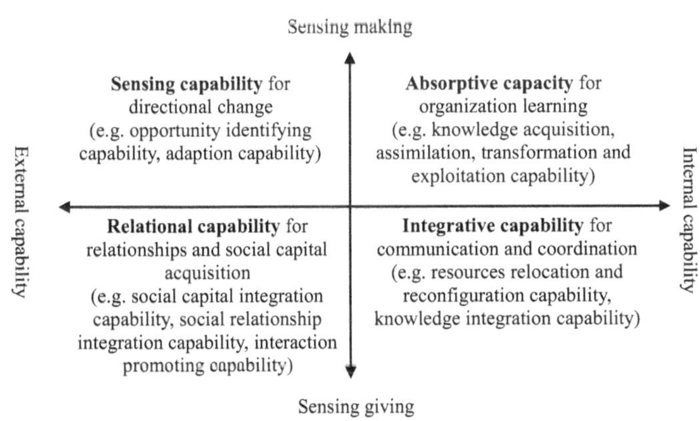

Fig. 6.2 Components of dynamic capabilities

of relocating, recombining and reusing both existing resources and those obtained, for example, resources relocation and reconfiguration capability, and knowledge-integration capability.

6.4 Hypotheses

The four dynamic capabilities identified in the literature may affect each phase in the process of adoptive management innovation as justified and hypothesized below.

6.4.1 Sensing Capability and Adoptive Management Innovation

The sensing capability refers to a subjective perception of environmental change; identifying and shaping opportunities (Teece 2007; Wang and Ahmed 2007). As Kor and Mahoney (2004), and Wiklund and Shepherd (2003) argue that, managers' perceptions are a major driver of firm-level heterogeneity and differential capacity. The subjectivist theory of entrepreneurship (Kor and Mahoney 2004; Kor et al. 2007) particularly highlights this subjective perception, and sensing capability in entrepreneurial discovery and creativity. Therefore, subjectivist theory is also relevant to management innovation as such entrepreneurship often includes innovation in organizational design, leadership, and financing (Daily et al. 2002): as does management innovation. From this subjectivist perspective, a given environment or situation does not strictly determine innovation decision-making alternatives and choices. Substantial possibilities occur for the evaluation, creativity and autonomy of individual choice (Penrose 1959) from initiation through to implementation. That is, innovative activities in organizations occur through managers' subjective sensing of processes that needs to be initiated to then search, learn, establish, and implement this creativity. Consistent with the subjectivist perspective, Schumpeter (1934) advocates that innovation, especially major organizational changes, depends on "intuition, the capacity of seeing things in a way which afterwards proves to be true" (Schumpeter 1934, p. 85); presumably, after implementation. Similarly, Kirzner (1973, p.109) argues that: "human action involves a posture of alertness toward the discovery of as yet unperceived opportunities and their exploitation"; a key activity of initiation and outside searching. Kor et al. (2007) further consider the development of intuitive innovation depends on managers' experiences and knowledge which are distributed, tacit and subjective.

Thus, proposals of management innovation differ between firms because managers have different perceptions of environmental changes, internal advantages, existing management solutions, and successfully implementing innovations. In other words, sensing capability of firms may affect the whole process of adoptive management innovation. For example, firms rely on sensing capability to identify a novel problem or perceive a shortfall between the organization's current and potential performance (Birkinshaw et al. 2008; Guillén 1994) to adopt an innovation in management practices. This initiation can be complicated to implement because of high uncertainty and potential opportunism (Casson and Godley 2007; Foss 1994; Foss et al. 2007; Witt 2007) but sensing capability may reduce the uncertainty and seize opportunities by foreseeing problems and proposing solutions through the different stages of the process. Essentially, management innovation facilitate sensing capability to motivate, invent, and implement (Gebauer 2011).

The purpose of management innovation is to adapt to the external environment changes at the same time to solve the problems within the organization, therefore, management innovation began on the external environment changes and internal problems of perception, that is, the sensing ability is the primary dynamic capacity to promote management innovation. In order to avoid deviations from the direction of the environment, the organization needs to be aware of its changes at any time and make adjustments or changes to the internal operation, which is the original intention of managing innovation. In order to better meet the new requirements of the community for environmental protection, Guangzhou Honda implemented environmental performance management model innovation, to enhance the efficiency of internal operations while maintaining the same pace with the external environment change, it was a win-win behavior of the organization. In addition to the external environment, the opportunity to management innovation also comes from perceiving changes of the organization's internal environment or problems. Besides, part of the direct observations or corporate statements can be intuitive responses to the inefficient, cost-effective, repeat operations and other issues, and more internal problems, which are hidden in organizational practices and systems or employees, such as employee dissatisfaction, departmental contradictions, low efficiency and organizational structures and mechanisms, are not suited to market demands. These issues need to be addressed by management innovation. Hence, the study here proposes that firms may rely on sensing capability to initiate evaluation of management practices, perform an outside search, sense potential risks and rewards of innovations in proposal establishment, perceive employees' attitudes towards adopting management practices, foresee progress and effectiveness of the innovation; to then make any necessary adjustments in innovation implementation.

Hypothesis 1a-d: Sensing capability of organizations positively influences performance of initiation, outside search, proposal establishment and implementation of adoptive management innovation, respectively.

6.4.2 Absorptive Capacity and Adoptive Management Innovation

Studies indicate an implicit consensus on defining the absorptive capacity as a set of abilities to manage knowledge (e.g. Tavani et al. 2014; Tu et al. 2006; Zahra and George 2002). Zahra and George (2002) conceptualize absorptive capacity as a dynamic capability pertaining to knowledge creation and utilization that enhances a firm's ability to gain and sustain a competitive advantage. Similarly, Cohen and Levinthal (1990, p. 128) define absorptive capacity as a firm's ability to "recognize the value of new information, assimilate it, and apply it to commercial ends". Researchers (e.g. Cheng and Chen 2013; Lewin et al. 2011; Volberda et al. 2010) even see absorptive capacity as a function of a firm's prior knowledge which is particularly related to how well the firm can use new knowledge to achieve desirable management innovation. Researchers further argue that the capabilities needed to manage the process of innovation could be developed from the absorptive capability (Davenport et al. 2006; Hertog et al. 2010). As Gassmann et al. (2010) and Rosenkopf and Nerkar (2001) suggest that, when absorptive capacity is embedded over time, organizational routines become more valuable, inimitable, and non-substitutable, which may increase a firm's ability to evaluate and use new management concepts, methods and skills in innovation (Zahra and George 2002). Thus, the firm is able to quickly identify innovation opportunities, experiment with implementation, and engage in management innovation (Gassmann et al. 2010; Rosenkopf and Nerkar 2001). Therefore, the accumulation of absorptive capacity is able to facilitate the whole process of complicated innovations such as organizational changes or adoptive management innovation.

Zahra and George (2002) suggest a multi-dimensional construct of absorptive capacity and propose four factors: acquisition refers to activities of identifying and initiating the knowledge acquisition that is critical to adoption of new management practices; assimilation refers to activities of analyzing, processing, interpreting, and understanding the information obtained (Kim1997; Szulanski 1996); transformation reflects the activities of combining existing knowledge and the newly acquired and assimilated knowledge; and exploitation refers to activities of creating further new knowledge (Tiemessen et al. 1997; Van den Bosch et al. 2003). These four components coexist and fulfill a necessary but insufficient condition for the whole process of management innovation (Birkinshaw et al. 2008; Cheng and Chen 2013) but they can build on each other to make absorptive capacity a coherent dynamic capability that fosters important innovation in management practices or methods. Specifically, initiating a management innovation relies on a firm's absorptive capacity to acquire and assimilate innovation knowledge; searching outside for management practices to propose relies on an absorptive capacity to quickly recognize, acquire and clearly understand valuable knowledge; reestablishing a new scheme relies on a process of value creation through knowledge transformation and exploitation; and implementation relies on exploitation of this knowledge (Lin and Su 2014).

Organizations that use adoptive management innovation are trying to learn from the practices or methods of other firms, and therefore need to assess, digest and utilize existing external management knowledge, that is, absorptive capacity. Absorptive capacity reflects the organization existing knowledge and knowledge structures in a certain extent. In general, the higher the level of absorptive capacity, the higher the firm's ability to learn and integrate external information and convert it into embedded knowledge. High-absorbing capacity enterprises can hold a more successful and effective introduction of new management ideas or methods to bring positive results to improve performance. However, the enterprises with low absorptive capacity tend to be hindered in obtaining and digesting new knowledge about management innovation, resulting in backward management and low efficiency, and ultimately affect the successful implementation of adoptive management innovation. In short, organizations with strong absorptive capacity can effectively grasp the dynamic information of the environment, innovation knowledge and the process information, transform the knowledge into the embedded knowledge of the organization, realize the internalization of knowledge and create new knowledge, so as to promote the realization of the whole management innovation process (Lin and Su 2012).

Hypothesis 2a-d: Absorptive capacity of organizations positively influences performance of initiation, outside search, proposal establishment and implementation of adoptive management innovation, respectively.

6.4.3 Integrative Capability and Adoptive Management Innovation

Integrative capability, focusing on readjusting, relocating and reconfiguring the organization's functional capabilities to obtain cooperative implementation with efficiency and flexibility, is one of the fundamental roles of dynamic capabilities (Liao et al. 2009; Teece 2007; Winter 2003; Zahra et al. 2006). Consistent with the combinative capability proposed by Kogut and Zander (1992) and the reconfiguring capability by Teece (2007), the integrative capability presents a process of complementing existing resources through enhancing, replicating and eliminating successfully (Helfat and Peteraf 2003). Briefly, integrative capability reflects a higher order of organizational capabilities in utilizing existing resources and capabilities to rebuild organizational routines and practices (Collis 1994).

Within the dynamic capability literature, two groups of integrative capability are identified: external-oriented studies deal with sensing, detecting, identifying, filtering, and calibrating opportunities; and internal-oriented studies deal primarily with seizing, capitalizing, and exploiting opportunities through intra-firm structures, procedures, designs, and incentives. Specifically, Liao et al. (2009) report a positive and significant effect of integrative capability on innovation by differentiating external integrative capability for opportunity recognizing from internal integrative capability for opportunity capitalizing. Since many characteristics of external integrative capability are measured in sensing capability and absorptive capacity, the study here focuses on internal integrative capabilities. The study argues that integrative capability is able to support the whole process of management innovation by continuously renewing and offering resources through integration. This can be likened to the process that typically arises in new technology innovation through novel combinations of existing ideas and practices (Kogut and Zander 1992). For example, initiating an innovation in management practices relies on the capability to integrate information sources from outside searches with those internally generated to identify novel problems whilst simultaneously integrating the interests and demands of various stakeholders; establishing a new proposal depends on the capability to integrate existing management practices with solutions proposed, existing knowledge bases and internal resources. Finally, implementation of management innovation depends on such capabilities of a firm to acquire, absorb, and assimilate internal and external sources of knowledge (Cohen and Levinthal 1990; Henderson and Clark 1990), integrate and reintegrate a firm's resource base (Teece et al. 1997), deploy and redeploy a firm's resources (Rumelt 1987), thereby rebuilding and reintegrating the well-entrenched organizational routines and practices that are unsustainable.

Hypothesis 3a-d: Integrative capability of organizations positively influences performance of initiation, outside search, proposal establishment and implementation of adoptive management innovation, respectively.

6.4.4 Relational Capability and Adoptive Management Innovation

Relational capability refers to a firm's ability to build close relationships and utilize resources in its network to realize goals of management innovation (Barreto 2010; Helfat et al. 2007). As with integrative capability, two types of relational capability are identified: external relational capability and internal relational capability. Externally, a firm as a system does not exist in isolation but is embedded in a network of relationships. Therefore, external relational capability is the result of a gradual process in which firms decide to broaden their vision of relationships by setting up a wide array of social and economic relationships with other organizations or individuals (for example, suppliers, trade association memberships, interlocking directorates, and prior strategic alliances), aiming to improve their competitive market position (Gulati and Gargiulo 1999; Webster 1992). Relational capabilities facilitate the ability to interact and share significant knowledge (Lorenzoni and Lipparini 1999) when searching outside. Internally, the internal network and all its members are those on which all activities of the firm depend. Internal relational capability initiates interaction, communication, knowledge and value sharing across all relationships with the firm. Factors such as commitment to relationships (Moorman et al. 1992), building trust and communication (Mohr and Spekman 1994) are widely recognized as being key determinants and the glue that binds high-quality external and internal relationships. These relationships benefit firms by providing opportunities for sharing financial resources (Ingram and Inman 1996; Keister 1998), institutional resources (Baum and Oliver 1991), as well as a host of other resources (Ingram and Inman 1996; Laursen and Salter 2006).

When a firm requires various resources to initiate and implement innovations in management practices, the firm often complements its current resources by accessing relationships (Adler and Kwon 2002; Cooper et al. 1995). Thus, the present study argues that the relational capability of a firm may facilitate the whole process of adoptive management innovation. For example, sensing opportunities is likely to be the result of communication within relationships, and outside searching for existing management practices relies on relational capabilities to obtain valuable information (Lin and Su 2014).The impact of relational capability on the process of management innovation is mainly reflected in the two stages of opportunity recognition and program creation, because these two stages are based on knowledge acquisition, especially external knowledge acquisition, and due to the limited knowledge of employees and managers, the diversification of organizational social networks becomes an important way to acquire knowledge. In other words, the organization through the interaction with the network members can achieve knowledge acquisition, and market knowledge source and professional knowledge source are important ways to obtain external knowledge, including advisory bodies, peer companies, non-peer enterprises, cooperative enterprises, suppliers and customers; professional knowledge sources are involved in experts, industry organizations or associations, national or local government agencies, etc., all of which are important compositions of organization social networks. Relational capability directly affects the effectiveness of the organization's external information, thus determining the organization's awareness of management innovation opportunities and the quality of innovative programs. In addition, in the management innovation implementation stage, the relational capability is mainly reflected in the organization's internal network, that is, through the interaction between the staff to achieve information exchange, form a common understanding of the new management practices, condense into a collective force, and jointly promote the development of innovative activities. In summary, organizational relational capability affects the entire process of management innovation, where the impact of opportunity identification and program creation decision-making is greater than that in the implementation phase (Lin and Su 2012). In addition to inside sources of knowledge, Mol and Birkinshaw (2009) suggest two outside sources that represent important relationships for firms seeking new management practices: market sources and professional sources. Having more ties with a variety of knowledge sources may positively affect the adoption of new practices (Mol and Birkinshaw 2009). In other words, firms can increase their chances of identifying existing management practices beneficial to them and gain insights into those that have been beneficial elsewhere by building more trust-based relationships and regularly searching this pool of knowledge. Studies also indicate that more relationships may facilitate knowledge acquisition by improving efficiency and cutting costs (De Carolis and Saparito 2006); heterogeneous relationships offer more knowledge sources to meet the demand for diversified knowledge of new management practices (Lin and Su 2014); and even abundant weak relationships are important in searching for a greater diversity of information (Granovetter 1973, 1985). Although firms may put considerable faith in external trusted and neutral professional peers when establishing a new innovation proposal that matches their specific context and implementing the proposal (Birkinshaw et al. 2008), a lack of a mature cooperation mechanism between firms and associations or professional institutions in China may lead to less efficiency of external relationships. Therefore, the study here argues that firms rely on the cohesive power of internal members (i.e., internal relational capability) to enhance the performance of proposal establishment and implementation by exchanging information internally and obtaining support for the innovation.

Hypothesis 4a-d: Relational capability of organizations positively influences performance of initiation, outside search, proposal establishment and implementation of adoptive management innovation, respectively.

6.4.5 The Fundamental Role of Relational Capability

The four components of dynamic capabilities are conceptually distinct and affect the adoptive management innovation through unique paths, but they are closely related rather than mutually exclusive. Particularly, sensing capability, absorptive capacity, and integrative capability all rely on information or resources acquisition through external and internal relationships, such that the relational capability of firms plays a fundamental role by setting up close relationships with other firms, institutions, individuals as well as internal employees. The importance of networking and alliances has been recognized (Gimeno 2004; Nohria and Garcia-Pont 1991) in appreciating the dynamics of relationships (Brass et al. 2004) and the need for further research of the role they play (Capaldo 2007).

Firms with sensing capabilities are adept at sensing a better opportunity for changes and innovations, and subsequently searching for solutions (Gulati and Gargiulo 1999). Whenever these firms sense a need for an outside search for such solutions, they tend to seek support through their established relationships. In other words, relational capabilities enable continuous interaction (internally and externally) to facilitate obtaining and sharing of significant information and resources (Ingram and Inman 1996; Laursen and Salter 2006). In a highly uncertain situation, relational capabilities are nurtured as the firm interacts more with external parties in order to gain access and absorb relevant knowledge and resources (Powell 1998, p. 229), and as employees of firms interact more with each other to share information. Singh and Zollo (2004) refer to forms of absorptive capabilities such as experiential learning and institutionalized learning and suggest that they are developed through relational capabilities, and subsequently integrative capabilities (for example, alliances and acquisitions). Similarly, empirical evidence from a study of 253 suppliers to the equipment industry finds that joint knowledge acquisitions has a positive effect on relational performance (Mesquita et al. 2008). Further, in a large-scale analysis of recent research on inter firm contracting, relational capabilities are found to be one of the key concepts affecting numerous aspects including enabling learning, knowledge acquisition, robustness, joint action, and trust (Schepker et al. 2014). Significantly, relational capability is also the ability to coordinate internal and external activities, manage conflict, foster trust and encourage information exchange through relationships, which enables knowledge acquisition and exploitation (i.e., two factors of absorptive capacity). Salunke et al. (2011) refer to this capability as relational learning capability, and conceptualize the capability as the process underpinning dynamic capabilities. Balbastre and Modeno-Luzon (2003) analyze this learning process within organizations and categorize the process as interactive relationships on three specific ontological levels: individual, group and organizational. Well-developed relational management skills afford firms the capability to pursue integrative agreements (McGrath and O'Toole 2008), especially internal integration of major changes such as management innovation. McGrath and O'Toole (2008) define relational capability as the capacity of a firm to interact proactively with a wide range of connected actors to purposefully exchange knowledge, create opportunities and joint process improvements including integration and innovations. Therefore, relational capability serves as a basis for other dynamic capabilities. Subsequently, the better the relational capability is developed, the better sensing capability, absorptive capacity and integrative capability work, which enables the innovation process from initiation through to implementation.

Hypothesis 5a: Relational capability of organizations positively influences their sensing capability.

Hypothesis 5b: Relational capability of organizations positively influences their absorptive capacity.

Hypothesis 5c: Relational capability of organizations positively influences their integrative capability.

In conclusion, the four phases of adoptive management innovation are affected by these four dynamic capabilities. Then, the study set up the structural model as appearing in Fig. 6.3, and tested the model through PLS-SEM by using Smart-PLS 2.0 software.

6.5 Method

6.5.1 Research Setting and Data Collection

The preliminary questionnaire was designed based on the relevant research literature and combined with practices of enterprises. Before conducting the survey, a questionnaire was designed in an iterative manner. Indicators for dynamic

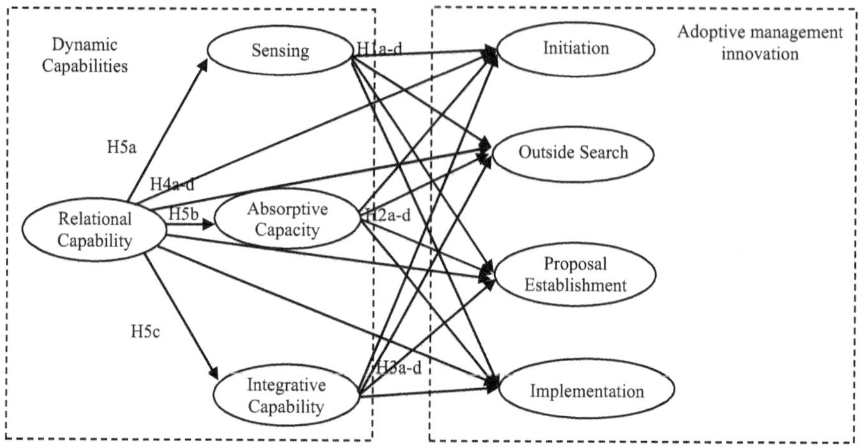

Fig. 6.3 The structural model

capabilities and adoptive management innovation were then identified respectively. Five peers of the author and five industry representatives were consulted, and a pilot study with 50 respondents was conducted. The sample consisted exclusively of manufacturing firms that tend to adopt similar management practices in China. An initial email was sent in March 2012 to managers of 1200 Chinese manufacturers with a minimum of 3 year tenure, and ensured their firms had adopted management innovation during the period by asking the question that "did your company make major changes in the following areas of business practices during 2009–2011 by adopting management practices? (1) strategies;(2) management techniques; (3) organizational structure; (4) business process reengineering; (5) organizational culture; (6) marketing strategies; (7) financial management methods; (8) human resources management; (9) others (0 = not used; 1 = used)". 823 useable responses were received from firms that had adopted at least one management innovation. In April 2012, a follow up mail invited these respondents to complete an online questionnaire. Respondents were ensured confidentiality and offered access to survey results. 264 of the 823 respondents completed the survey, with a response rate of 32.1%. Respondents had an average tenure of 5.23 years (S.D. = 2.36) with an average company size of 235 employees (S.D. = 3.75).

Non-response bias. The study here used a T-test to examine differences between respondents and non-respondents. Results showed no significant differences ($p < 0.05$). A comparison between early and late respondents also did not reveal any significant difference ($p < 0.05$). Therefore, the sample was free of non-response bias.

Common method bias. Common method bias, a major validity threat in behavioral research (Podsakoff et al. 2003), was assessed through several steps. When designing the survey, the study chose clear and concise items, used different response formats and scale endpoints for the independent and dependent variables, and conducted scale-length control by following the recommendations of Podsakoff et al. (2003). When surveying, the study assured respondents of confidentiality and encouraged honest responses by reminding them that there were no right or wrong answers. The study also improved scale items by consulting both academic experts and industry representatives. Following data collection, Harman's one-factor test was conducted to assess common method bias. Eight factors were extracted, accounting for 60.534% of the variance explained. Since "this procedure actually does nothing to statistically control for (or partial out) method effects" (Podsakoff et al. 2003, p. 889), the study further controlled for the effect of a single unmeasured latent method factor (Podsakoff et al. 2003; Vaccaro et al. 2012). To do so, two confirmatory factor analysis models were constructed: in the first one, all items were allowed to load on their theoretical factors; while in the other one, all items were allowed to load on a latent common factor. A comparison was then made and the result indicated a better fit in the latter model (χ^2 = 104.36, df = 48) over the theoretical model (χ^2 = 122.41, df = 52), and the latent common factor accounted for only a very small portion of the total variance. Therefore, the common method bias was not a pervasive problem here.

6.5.2 *Measures*

Dependent Variable. As discussed above, a process-oriented measurement was suggested in the study. First, the study measured four phases of management innovation, extracted the characteristics of each phase, and explored how adoptive innovation was different from the generating innovation. A pool of items tapped into different phases of management innovation for the initial questionnaire. New items were added and others deleted through the consult. Finally, the phrasing of the items was further validated, and a final version of adoptive management innovation with 16 items was developed, (item 1 to 4 for initiation, item 5 to 8 for outside search, item 9 to 12 for proposal establishment, and item 13 to 16 for innovation implementation) as appearing in Appendix. All items were measured on a 5-point Likert scale, on which "1 = strongly disagree' and '5 = strongly agree".

Independent Variable. The study measured each dimension of dynamic capabilities by using existing research and scales. Respondents rated items of the four dimensions on a 5-point scale. More specifically, the construct of sensing capability was operationalized from previous studies such as Wang and Ahmed (2007) and Wu et al. (2012) as SC-1 to SC-5 (i.e., experience and knowledge of managers, the number of paths for transferring information on external changes, the number of paths for transferring information from bottom up, information selecting, new opportunities detecting and reacting); the absorptive capacity was based on research of Jansen et al. (2005) and Zahra and George (2002) as AC-1 to AC-4 (i.e., knowledge acquisition, assimilation, transformation, and exploitation); the integrative capability was based on Kogut and Zander (1992) and Liao et al. (2009) as IC-1 to IC-4 (i.e., resources identification, a system for resources transferring, the effectiveness of vertical and horizontal communication and cooperation); the relational capability was based on studies of Adler and Kwon (2002) and Subramaniam and Youndt (2005) as RC-1 to RC-5 (i.e., an effective external network with a central position, a close relationship with the government and institutions, cooperating relationships with other firms, an effective internal network, and a trust-based internal network).

6.5.3 The Measurement Model

The study then set up the measurement model, shown in Fig. 6.4. Since all indicators were reflective, there was no need to change the arrow direction (Hair et al. 2013; Sarstedt et al. 2014).

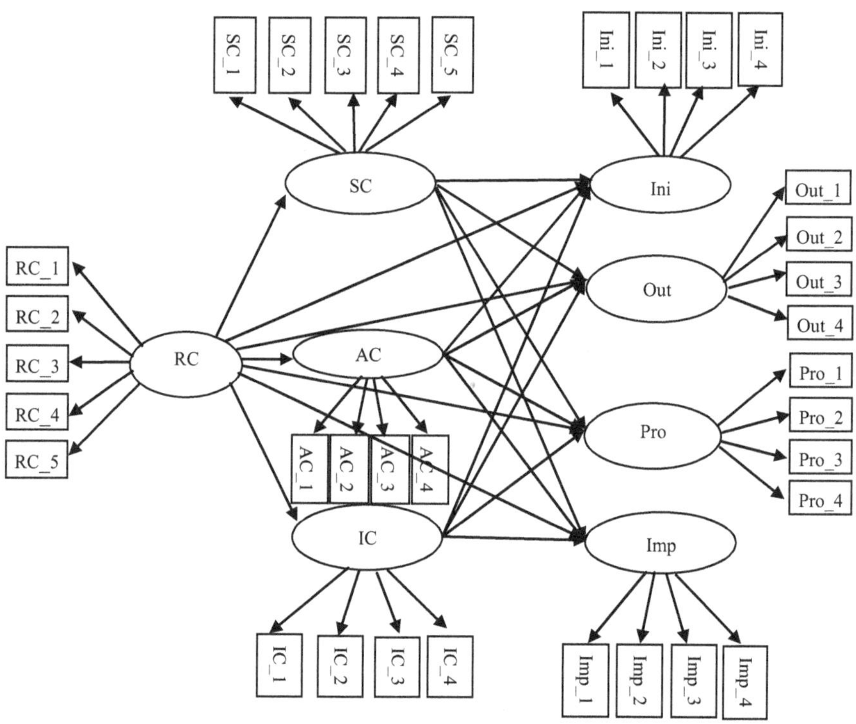

Fig. 6.4 The measurement model. *Note* RC relational capability, SC sensing capability, AC absorptive capacity, IC integrative capability, Ini initiation, Out outside search, Pro proposal establishment, Imp implementation

6.6 Analysis and Results

Structural Equation Modeling (SEM) is able to simultaneously estimate and test the above models with collected data. As Hair et al. (2012, 2013) argue, Partial Least Squares SEM (PLS-SEM) has become a good alternative to Covariance-Based SEM (CB-SEM) for estimating theoretically justified cause-effect relationship models especially when the sample size is small. The study here adopted PLS-SEM and used Smart-PLS 2.0 by Ringle et al. (2005). Following PLS-SEM guidelines (e.g. Hair et al. 2012, 2013; Gudergan et al. 2008; Reimann et al. 2010; Sarstedt et al. 2014), the study performed a two-stage approach to evaluation: (1) assessment of measurement model (outer model), and (2) estimation of structural model (inner model) and hypothesis tests.

6.6.1 Assessment of the Measurement Model

All constructs drew on a reflective measurement model in the study, because the indicators of each constructs are correlated and interchangeable (Hair et al. 2013). The study conducted Stage 1 by assessing reliability and validity of constructs. Results are reported in Table 6.1.

Indicator Reliability. Table 6.1 shows that all indicators have individual indicator reliability values (i.e., loading^2) that are greater than the minimum acceptable level of 0.4 and close to the preferred level of 0.7, except for Out_1 for Outside search (0.182) and Pro_2 for Proposal establishment (0.305) which had adverse effects on the construct measures' convergent validity and internal consistency reliability and so were deleted.

Internal consistency reliability. Advocators of PLS-SEM (e.g. Hair et al. 2012, 2013; Sarstedt et al. 2014) suggest "composite reliability" as a replacement for "Cronbach's alpha" in assessing internal consistency reliability. According to Hair et al. (2013), values between 0.60 and 0.70 are considered "acceptable in exploratory research", whereas values between 0.70 and 0.95 are "satisfactory to good", and values above 0.95 are problematic. From Table 6.1, all values are >0.6: demonstrating internal consistency reliability.

Convergent validity. Convergent validity measures the extent a construct converges in its indicators by explaining the items' variance by the Average Variance Extracted (AVE). Results indicated all construct AVE values were greater than the acceptable threshold of 0.5 suggested by Hair et al. (2013).

Discriminant validity. Discriminant validity determines the extent to which a construct is empirically distinct from other constructs in the path model. As Fornell and Larcker (1981) suggest, the square root of AVE in each latent variable can establish discriminant validity, if this value is greater than the correlation values with all other latent variables. The correlation matrix in Table 6.2 shows that discriminant validity was thus established for all constructs.

Table 6.1 Results Summary for reflective outer models

Variables	Indicators	Loadings	Indicator reliability	Composite reliability	AVE
Relational capability	RC_1	0.824	0.679	0.844	0.741
	RC_2	0.852	0.726		
	RC_3	0.879	0.773		
	RC_4	0.891	0.794		
	RC_5	0.857	0.734		
Sensing capability	SC_1	0.851	0.724	0.824	0.764
	SC_2	0.903	0.816		
	SC_3	0.826	0.682		
	SC_4	0.873	0.762		
	SC_5	0.915	0.837		
Absorptive capacity	AC-1	0.838	0.702	0.873	0.742
	AC-2	0.865	0.748		
	AC-3	0.891	0.794		
	AC-4	0.852	0.726		
Integrative capability	IC-1	0.843	0.711	0.801	0.691
	IC-2	0.822	0.676		
	IC-3	0.827	0.684		
	IC-4	0.832	0.692		
Initiation	Ini_1	0.872	0.760	0.789	0.744
	Ini_2	0.852	0.726		
	Ini_3	0.843	0.711		
	Ini_4	0.883	0.780		
Outside search	Out_2	0.921	0.828	0.862	0.781
	Out_3	0.935	0.874		
	Out_4	0.824	0.679		
Proposal establishment	Pro_1	0.858	0.736	0.851	0.784
	Pro_3	0.904	0.817		
	Pro_4	0.898	0.806		
Implementation	Imp_1	0.861	0.741	0.860	0.742
	Imp_2	0.879	0.773		
	Imp_3	0.880	0.774		
	Imp_4	0.825	0.681		
	Out_1	0.426	0.182	$Loading^2 < 0.4$, deleted	
	Pro_2	0.552	0.305	$Loading^2 < 0.4$, deleted	

Table 6.2 Correlation of latent constructs

	Relational capability	Sensing capability	Absorptive capacity	Integrative capability	Initiation	Outside search	Proposal establishment	Implementation
Relational capability	0.861							
Sensing capability	0.548	0.874						
Absorptive capacity	0.625	0.593	0.861					
Integrative capability	0.582	0.612	0.532	0.831				
Initiation	0.482	0.598	0.563	0.552	0.863			
Outside search	0.471	0.423	0.662	0.497	0.526	0.884		
Proposal establishment	0.315	0.482	0.431	0.482	0.428	0.386	0.885	
Implementation	0.442	0.324	0.372	0.437	0.625	0.426	0.529	0.861

6.6.2 Estimation of the Structural Model

After establishing reliability and validity, the study assessed the structural model by re-sampling about 200 times to reach the number of 5000 samples for bootstrapping (Hair et al. 2013).

Collinearity. Variance Inflation Factor (VIF) values were used to examine the collinearity. The result showed that VIF values for all independent variables ranged between 1.436 (sensing capability) and 2.442 (integrative capability), indicating that the results were not negatively affected by collinearity as they were all <5 (Hair et al. 2013).

Coefficient of determination (R^2). The R^2 value of each endogenous construct is a measure of the variance explained in each endogenous construct and the model's predictive accuracy. According to Hair et al. (2013) and Sarstedt et al. (2014), R^2 values of 0.75, 0.50 and 0.25 may be considered substantial, moderate and weak, respectively. Results for initiation, outside search, proposal establishment, and implementation had moderate R^2 values of 0.575, 0.468, 0.403 and 0.512, respectively. Sensing capability, absorptive capacity, and integrative capability had comparably weak R^2 values of 0.242, 0.321, and 0.254, respectively. However, considering the possibility of extrinsic factors and alternatives, their R^2 values are satisfactory.

Cross-validated redundancy (Q^2). A Stone-Geisser's Q^2 value greater than zero for any endogenous construct indicates that the predictive relevance of the paths in the model is acceptable(Hair et al. 2013). Running the blindfolding procedure with an omission distance of eight yielded cross-validated redundancy values for all seven endogenous constructs well above zero (absorptive capacity: 0.342, sensing capability: 0.422, integrative capability: 0.263, initiation: 0.224, outside search: 0.287, proposal establishment: 0.364, and implementation: 0.450), supporting the model's predictive relevance.

The path coefficients. Figure 6.5 shows the results from the bootstrapping procedure (264 cases, 5000 samples, no sign changes option), and Table 6.3 presents direct and total effects of the four dynamic capabilities on the four phases of adoptive management innovation.

The R^2 for the dependent variables of initiation, outside search, proposal establishment, and implementation of adoptive management innovation indicated the four dynamic capabilities explained 57.5, 46.8, 40.3, and 51.2% of the variance in the dependent variables, respectively.

The R^2 values for sensing capability, absorptive capacity, and integrative capability indicated that relational capability explained 24.2, 32.1, and 25.4% of their variance. Although these percentages were relatively low, bootstrapping results revealed positive and significant effects of relational capability on the other three capabilities. Relational capability had the strongest effect on absorptive capacity ($\beta = 0.310$, $p < 0.01$), followed by integrative capability ($\beta = 0.286$,

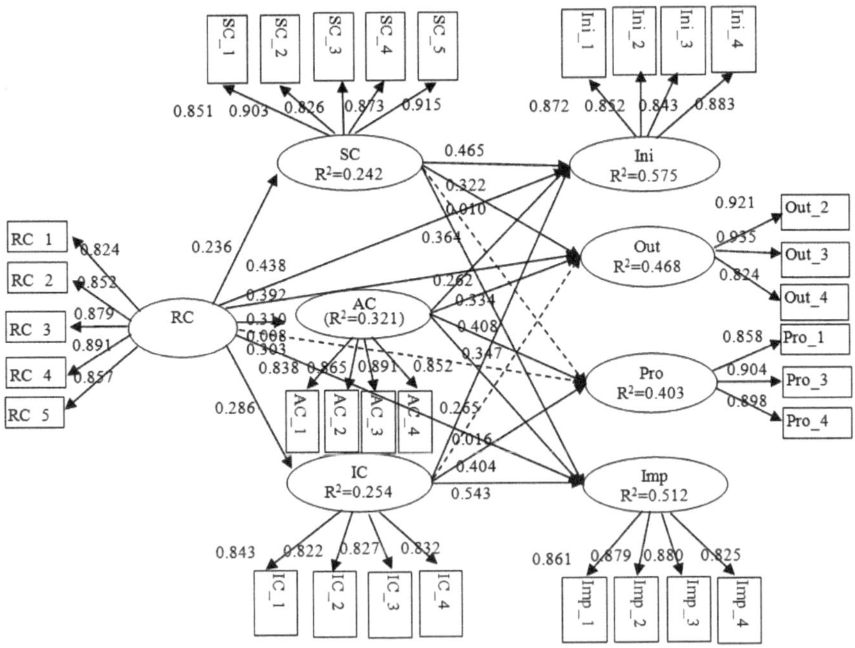

Fig. 6.5 PLS path coefficients and bootstrap statistics. *Note* RC relational capability, SC sensing capability, AC absorptive capacity, IC integrative capability, Ini initiation, Out outside search, Pro proposal establishment, Imp implementation

Table 6.3 The direct and total effects

Relationship	Direct effect	Total effect
Sensing capability → Initiation	0.465***	0.465***
Sensing capability → Outside Search	0.322***	0.322***
Sensing capability → Proposal establishment	0.010	0.010
Sensing capability → Implementation	0.364***	0.364***
Absorptive capacity → Initiation	0.262***	0.262***
Absorptive capacity → Outside Search	0.334***	0.334***
Absorptive capacity → Proposal establishment	0.408***	0.408***
Absorptive capacity → Implementation	0.347***	0.347***
Integrative capability → Initiation	0.265***	0.265***
Integrative capability → Outside Search	0.016	0.016
Integrative capability → Proposal establishment	0.404***	0.404***
Integrative capability → Implementation	0.543***	0.543***
Relational capability → Initiation	0.438***	0.705***
Relational capability → Outside Search	0.392***	0.576***
Relational capability → Proposal establishment	0.008	0.248***
Relational capability → Implementation	0.303***	0.652***

***$p < 0.01$

$p < 0.01$) and sensing capability ($\beta = 0.236$, $p < 0.01$). Therefore, H5a, H5b and H5c were supported.

Sensing capability had the strongest and positive effect on initiation ($\beta = 0.465$, $p < 0.01$), followed by implementation ($\beta = 0.322$, $p < 0.01$) and outside search ($\beta = 0.364$, $p < 0.01$). H1a, H1b and H1d were supported. The relationship between sensing capability and proposal establishment was not significant ($\beta = 0.01$, $p > 0.1$), so H1c was not supported.

Results showed the importance of absorptive capacity in all four phases of innovation, with the strongest effect on proposal establishment ($\beta = 0.408$, $p < 0.01$), followed by implementation ($\beta = 0.347$, $p < 0.01$), outside search ($\beta = 0.334$, $p < 0.01$) and initiation ($\beta = 0.262$, $p < 0.01$), thus supporting H2.

Results showed the importance of integrative capability on implementation ($\beta = 0.543$, $p < 0.01$), followed by proposal establishment, ($\beta = 0.404$, $p < 0.01$), and initiation ($\beta = 0.265$, $p < 0.01$), thus supporting H4a, H4c, and H4d. However, the relationship between integrative capability and outside search was not statistically significant ($\beta = 0.016$, $p > 0.1$), so H4b was not supported.

The effects of relational capability on initiation, outside search and implementation were positive and significant ($\beta = 0.438$, $p < 0.01$; $\beta = 0.392$, $p < 0.01$; $\beta = 0.303$, $p < 0.01$, respectively). Relational capability did not have a significant direct effect on proposal establishment ($\beta = 0.008$, $p > 0.1$), but did have a significant total effect ($\beta = 0.248$, $p < 0.01$). Most significantly, the total effects of relational capability on each phases were stronger than the direct effects; initiation ($\beta = 0.705$, $p < 0.01$), followed by implementation ($\beta = 0.652$, $p < 0.01$), outside search ($\beta = 0.576$, $p < 0.01$), and proposal establishment ($\beta = 0.248$, $p < 0.01$). Hence, H3a, H3b, H3c and H3d were supported.

6.7 Discussion and Implications

6.7.1 Discussion

The study here focuses on the nature of adoptive management innovation, its four-phase process, and effects of dynamic capability's four components on each phase. The study links to several research areas and makes various conclusions by addressing the two research questions postulated in the introduction.

(1) *The focus on adoptive management innovation and a process-oriented measurement*

In contrast to the focus of Birkinshaw et al. (2006, 2007, 2008) on new-to-the-state-of-the-art management innovation, and in consistence with Lin and Su (2014), Mol and Birkinshaw (2009), and Vaccaro et al. (2012), the research here focuses on new-to-the-organization management innovation, namely adoptive management innovation, through micro-level organizational behaviors. As the majority of firms (particularly, less-developed Chinese firms) prefer this lower-risk form of innovation (Lin and Su 2014), the research may be considered more reflective of current management practices.

Although other researchers advocate a result-oriented method for measuring management innovation (e.g. Mol and Birkinshaw 2009; Vaccaro et al. 2012), they also acknowledge the results are intangible, uncertain, lagging and even inseparable from that of technological innovation (Birkinshaw et al. 2008), and therefore difficult to operationalize. Building on other frameworks of management innovation (e.g. Birkinshaw et al. 2007, 2008; Hamel 2006; Lin and Su 2014), and a case study from Haier Group, China, the study designs a four-phase model consisting of initiation, outside search, proposal establishment, and implementation. Consistent with Lin and Su (2014), the four-phase framework attempts to demonstrate the differences between adoptive innovation and generating innovation where outside search and proposal establishment facilitate adoption to the idiosyncratic context of a firm. However, whilst Lin and Su (2014) take a rational perspective which focuses on roles of internal agents and even their mental activities; this study takes an organizational perspective by focusing on general activities taken by the firm. Moreover, Lin and Su (2014) regard the process of adoptive management innovation as two phases of adoption decision and implementation; the study clarifies each phase of the innovation process and effects of dynamic capabilities.

The reliability and validity assessment of the four-phase-process-oriented method showed that 14 of 16 items had reliability values >0.7, confirming a well-developed scale. The 14-item scale may further research on adoptive management innovation, and may also offer valuable insights for measuring generating innovation and other complex organization activities.

(2) *Effects of dynamic capabilities on each phase of adoptive management innovation*

Consistence with most research on the relationship between dynamic capabilities and innovation (e.g. Ambrosini et al. 2009; Aragon-Correa and Sharma 2003; Cheng and Chen 2013; Clausen 2013; Hart and Dowell 2010; Helfat et al. 2007; Kohlbacher 2013; Ridder 2011; Teece 2007; Teece et al. 1997), this research confirms the positive effects of dynamic capabilities on the process of adoptive management innovation by advocating a multi-dimensional construct view of a dynamic capability (e.g. Barreto 2010; Teece 2007; Wang and Ahmed 2007). Teece (2007) identifies three components from a macro perspective: sensing, seizing and reconfiguring; while the study identifies four components from both a macro and micro perspective of dynamic capabilities: sensing capability, absorptive capacity, relational capability, and integrative capability. This perspective may offer valuable insights for further research on dynamic capabilities, particularly at a micro level of analysis.

PLS SEM was used to examine both the structural model reflecting the relationships between dynamic capabilities and adoptive management innovation and the measurement model reflecting indicators (items) of each construct. As hypothesized, the effects of dynamic capabilities were generally confirmed, though not completely. Thus, the study here confirms dynamic capabilities as an important determinant in the management innovation process.

First, the study reveals positive and significant direct effects of sensing capability on three of the four innovation phases, especially initiation. This is consistent with the subjectivist theory which postulates the subjective perception and sensing of discovery (Kor and Mahoney 2004; Kor et al. 2007). The positive effect of sensing capability on initiation is also consistent with Birkinshaw et al. (2008) who suggest that the demand for management innovation is driven by identification of a novel problem or a perceived shortfall. When searching for existing management practices, being aware of their relative advantage, complexity, observability and compatibility, and seeking to gain a better understanding, firms depend on sensing capability for subjective evaluation and decision making. In addition, the positive effect of sensing capability on implementation demonstrates that managers depend on this capability throughout the process from initiation to implementation to make adjustments where necessary. Managers also sense employees' attitude towards adopted innovations, enabling them to negate adverse reactions in a timely manner. Since the results of management innovation are unlikely to be realized in the short term (Birkinshaw et al. 2008; Teece et al. 1997), managers may also need to sense its effectiveness, seek to draw experience and build a logical rationale for the link between the problem identified and the innovation proposed. However, the effect of sensing capability on proposal establishment is not significant, which indicates that firms may not revise innovation proposals by sensing changes. This may be led by cognitive biases (Bazerman 1990; Busenitz and Lau 1996) over managers' subjective perceptions. That is, managers tend to overestimate the reliability and accuracy of their perceptions when establishing an innovation proposal, or underestimate the hazards (De Carolis and Saparito 2006), and subjectively lower the level of risk perception.

Second, consistent with most pertinent studies (e.g. Cheng and Chen 2013; Davenport et al. 2006; Gassmann et al. 2010; Hertog et al. 2010; Rosenkopf and Nerkar 2001), the study here reveals positive effects of absorptive capacity on the whole process of innovation, especially proposal establishment. The study also furthers Lin and Su's (2014) study showing adoptive management innovation as a process of knowledge acquisition, assimilation, transformation and exploitation. More specifically, absorptive capacity may enable a firm to initiate a management innovation, efficiently obtain outside information, and clearly understand its potential value through knowledge acquisition and assimilation. Consequently, the firm is able to integrate innovations with internal resources through knowledge transformation and exploitation; and effectively implement innovations through knowledge exploitation.

Third, consistent with most research on integrative capability (e.g. Collis 1994; Teece 2007; Wang and Ahmed 2007; Winter 2003; Zahra et al. 2006), the study confirms the positive and significant effects of integrative capability on the innovation process, particularly implementation. This indicates a process that depends on a higher order of organizational capability in complementing existing resources through enhancing, replicating and eliminating successfully (Helfat and Peteraf 2003), or in utilizing existing resources and capabilities to rebuild organizational routines and practices (Collis 1994). More specifically, the study confirms the reliance of integrative capability on implementation and the essential task of setting up a new internal resource base for innovation, and supports Birkinshaw et al. (2008) and Lin and Su (2014) who suggest implementation as a process of integration. Moreover, implementation of management innovation depends on a recombination of resources. Only when resources are dynamically integrated to create synergy can optimal efficiency be realized. This internal synergy also enables all stakeholders to reach an agreement. The study also reveals that: firms depend on capability of integrating information on external environmental changes with those internally generated to identify novel problems and then initiate an innovation; and capability of integrating outside management practices with problems identified, existing knowledge base and internal resources to establish a new proposal. Liao et al. (2009) argue that integrative capability is able to foster the firm's awareness of innovation potentials and enhance their likelihood of creating innovative practices, the study here, however, sees integrative capability as a process of resources relocation and reconfiguration by taking an internal-oriented perspective. This may be the cause for the insignificant effect of integrative capability on outside search.

Finally, consistent with Ingram and Inman (1996), Keister (1998), Laursen and Salter (2006), and Lorenzoni and Lipparini (1999) who argue that relational capability positively affects innovation by facilitating the formation of the ability to interact and share significant knowledge and resources, the study reveals positive and significant total effects of relational capability on all four phases of adoptive management innovation. The study also confirms the view of Adler and Kwon (2002) and Cooper et al. (1995) that firms often complement their holding resources by accessing their relationships when initiating and implementing innovations. As a result of a gradual process in which firms extend the breadth of external relationships (Webster 1992), relational capability is able to stimulate existing knowledge acquisition and knowledge development (Powell 1998), and thus supporting the process of management innovation. However, external relational capability may produce different effects from internal relational capability. Initiation of innovation relies on external relational capability to efficiently obtain information on environmental changes and internal relational capability to obtain information on internal problems. Outside search depends mainly on external relational capability to facilitate knowledge transfer, which is consistent with Mol and Birkinshaw (2009) who argue firms can increase identification of existing innovations and insights by building trust with more knowledge outside sources (i.e., external relationships). The positive indirect effects of relational capability indicate that relational capability may affect the innovation process through the other three capabilities by interacting and sharing significant knowledge (Lorenzoni and Lipparini 1999). However, firms rely on internal relational capability to build coherent power among all stakeholders to implement innovations.

Additionally, the study also confirms relational capability as a fundamental driver of the other three capabilities. This may reflect the unique Chinese culture whereby favoritism or Guanxi rooting in networking leads to successful acquisition of resources and knowledge on which the formation of other capabilities relies. Building such trust-based relationships with other organizations, institutions, individuals and employees to obtain valuable information and support may determine the survival and development of organizations (Luo et al. 2013).

6.7.2 Managerial Implications

Adopting a PLS-SEM analysis based on 264 firms from China and blending management innovation and dynamic capability literature enabled development and examination of a relationship framework that explicated effects of dynamic capability on each phase of adoptive management innovation. This work offers a number of important insights for practices.

The process-oriented measuring for adoptive management innovation may also offer some valuable insights for firms in assessing system-dependent changes whose outputs tend to be intangible, uncertain, lagging and even inseparable. With difficulty in quantifying outputs of organizational practices, a process perspective may be adopted in assessing their performance. Furthermore, as the research indicates, the number of new practices initiated, attention paid to renewing management practices, management problem identification and timing of initiation are good indicators for performance of initiation; acquaintance with emerging practices in market sources, internal sources and professional sources are good indicators for performance of outside search; the preference on creating a new proposal, match of new established proposal with the original one and quality of proposals are good indicators for performance of proposal establishment; and the ratio of proposals taken into practice, the support of the whole organization, progress of implementation and outputs of innovation are good indicators for innovation implementation (with loadings all over 0.7).

How to improve the performance of management innovation has been a problem bothering managers of firms for a long time. This research offers a dynamic capability perspective by exploring and confirming effects of its four components on each phase of adoptive management innovation. The results indicate that management innovation is not isolated, but a systematic project depending on all the four components of dynamic capability.

More specifically, strong sensing capability, which could be reflected by indicators of experienced and knowledgeable managers, the number of paths for transferring information on external changes, the number of paths for transferring information up forward, quick selection of information, quick detection of new opportunities and making reactions (with loadings all over 0.7), may be able to directly or indirectly stimulate activities of outside search (with the strongest total effect), initiation and implementation. However, sensing capability is not able to improve performance of proposal establishment. Similarly, knowledge identification and acquisition, knowledge analysis and process, knowledge combination, and new knowledge creation are four important indicators of absorptive capacity (with loadings all over 0.7), which offers manager four paths to improve absorptive capacity of a firms. This research confirms positive and significant total effects of absorptive capacity on all four phases of innovation, especially a total effect of 0.677 on implementation, which reminds managers of the necessity of offering employees with knowledge acquiring, processing, combining and creating guidelines and tools. Integrative capability, with good indicators of resources identification, a system for resources transferring, effectiveness of vertical and horizontal communication and cooperation (with loadings all over 0.7), may be able to directly or indirectly support activities of implementation (with the strongest total effect of 0.693), proposal establishment and initiation. However, integrative capability does not significantly impact outside search due to its weak effect in the linkage. Finally, relational capability, with important indicators of an effective external network with a central position, a close relationship with the government and institutions, effectively cooperating relationships with other firms, an effective internal network with all employees involved and actively interacted, and a trust-based internal network, was confirmed as an important indicator of other three capacities. It may be the fundamental driver that adoptive management innovation depending on. Its effects on all four phases of innovation are positive and significant, especially with a strongest total effect on implementation of 0.926. Though it may not be able to produce direct effect on proposal establishment, it does affect proposal establishment through other capabilities or previous phases of innovation. As a result, more attention should be paid to set up close and trusted relationships with other firms, government and institutions and individuals as well as internal employees to support process of complicate organizational changes like management innovation to finally improve organizational performance.

6.8 Conclusion

Adopting management innovation from external firms is attractive to smaller and less complex firms. Reasons include limited resources, and ease of imitation without concern for patent violations. Further, this form of management innovation is a preferred option for many Chinese firms. However, use of limited yet valuable resources is insufficient to maintain a sustainable management innovation process: recurring dynamic capabilities are essential throughout for such effectiveness and efficiency in converting inputs to outputs.

In contrast to technological innovation, management innovation tends to be intangible and tacit in nature, with lagging and inseparable results. Such characteristics highlight the importance of measuring management innovation through a process-based scale. Focusing on adoptive management innovation, this research sets up a four-phase framework and develops a process-oriented measurement scale. Particularly, the study builds a framework between four distinct but related dynamic capabilities and four phases of adoptive management innovation, and examines the framework with PLS-SEM. The study demonstrates the importance of both an external and internal perspective, and highlights the importance of different capabilities at different stages of the innovation process. These capabilities may be characteristics of the personnel but should also be embedded, maintained, and updated in the firm's processes. The findings offer many valuable insights for further research of innovation, particularly in a Chinese context, with important implications for management practices.

Despite its limitations, the present study makes several important contributions to the emerging literature on management innovation. The study furthers Birkinshaw et al.'s (2008) study which focuses on generating innovation by extending to adoptive management innovation which is more popular in China and possibly globally. The study also builds on Lin and Su's (2014) research on adoptive management innovation by specifying the process of innovation and developing a new measurement scale. Further, the study combines a dynamic capability perspective with a management innovation perspective, and confirms that four dynamic capabilities positively affect each phase of innovation. In conclusion, the study explores the complex process of adoptive management innovation and illustrates the role of dynamic capabilities in the process; offering an effective approach for improving the performance of innovation.

References

Adler, P., Kwon, S.: Social capital: prospects for a new concept. Acad. Manag. Rev. **27**(1), 17–40 (2002)

Ambrosini, V., Bowman, C., Collier, N.: Dynamic capabilities: an exploration of how firms renew their resource base. Br.J. Manag. **20**(Supplement s1), S9–S24 (2009)

Aragón-Correa, J.A., Sharma, S.: A contingent resource-based view of proactive corporate environmental strategy. Acad. Manag. Rev. **28**(1), 71–88 (2003)

Armbruster, H., Bikfalvi, A., Kinkel, S., Lay, G.: Organizational innovation: the challenge of measuring non-technical innovation in large-scale surveys. Technovation **28**(10), 644–657 (2008)

Balbastre, F., Moreno-Luzón, M.: Self-assessment application and learning in organizations: a special reference to the ontological dimension. Total Qual. Manag. Bus. Excell. **14**(3), 367–388 (2003)

Barney, J.B.: Firm resource and sustained competitive advantage. J. Manag. **17**(1), 99–120 (1991)

Barreto, I.: Dynamic capabilities: a review of past research and an agenda for the future. J. Manag. **36**(1), 256–280 (2010)

Baum, J.A.C., Oliver, C.: Institutional linkages and organizational mortality. Adm. Sci. Q. **36**(3), 187–218 (1991)

Bazerman, M.: Judgment in Managerial Decision Making. Wiley, New York (1990)

Birkinshaw, J., Crainer, S., Mol, M.: Special report on management innovation. Bus. Strategy Rev. **18**(1), 45–73 (2007)

Birkinshaw, J., Hamel, G., Mol, M.J.: Management innovation. Acad. Manag. Rev. **33**(4) (2008)

Brady, T., Davies, A.: Building project capabilities: from exploratory to exploitative learning. Organiz. Stud. **25**(9), 1601–1621 (2004)

Brass, D.J., Galaskiewicz, J., Greve, H.R., Tsai, W.: Taking stock of networks and organizations: a multilevel perspective. Acad. Manag. J. **47**(6), 795–817 (2004)

Busenitz, L.W., Lau, C.M.: A cross-cultural cognitive model of new venture creation. Entrep. Theory Pract. **20**(4), 25–39 (1996)

Capaldo, A.: Network structure and innovation: the leveraging of a dual network as a distinctive relational capability. Strateg. Manag. J. **28**(6), 585–608 (2007)

Casson, M., Godley, A.: Revisiting the emergence of the modern business enterprise: entrepreneurship and the singer global distribution system. J. Manage. Stud. **44**(7), 1064–1077 (2007)

Cheng, C.C.J., Chen, J.S.: Breakthrough innovation: the roles of dynamic innovation capabilities and open innovation activities. J. Bus. Ind. Mark. **28**(5), 444–454 (2013)

Clausen, T.H.: The role of operational and dynamic capabilities in ambidextrous innovation: a longitudinal analysis. 35th DRUID Celebration Conference, Barcelona, Spain, 17–19 June 2013

Cohen, M.D., Bacdayan, P.: Organizational routines are stored as procedural memory: evidence from a laboratory study. Organ. Sci. **5**(4), 554–568 (1994)

Cohen, W.M., Levinthal, D.A.: Absorptive capacity: a new perspective on learning and innovation. Adm. Sci. Q. **35**(1), 128–152 (1990)

Collis, D.J.: How valuable are organizational capabilities? Strateg. Manag. J. **15**(8), 143–152 (1994)

Cooper, A.C., Folta, T.B., Woo, C.Y.: Entrepreneurial information search. J. Bus. Ventur. **10**(2), 107–120 (1995)

Daily, C.M., Mcdougall, P.P., Covin, J.G., Dan, D.R.: Governance and strategic leadership in entrepreneurial firms. J. Manag. **28**(3), 387–412 (2002)

Davenport, T., Leibold, M., Voelpel, S.: Strategic Management in The Innovation Economy, Strategy Approaches and Tools for Dynamic Innovation Capabilities. Wiley, New York, NY (2006)

De Carolis, D.M., Saparito, P.: Social capital, cognition, and entrepreneurial opportunities: a theoretical framework. Entrep. Theory Pract. **30**(1), 41–56 (2006)

DiMaggio, P.J., Powell, W.W.: The iron cage revisited: institutional isomorphism and collective rationality in organizational fields. Am. Sociol. Rev. **48**(2), 147–160 (1983)

Eisenhardt, K., Martin, J.: Dynamic capabilities: what are they? Strateg. Manag. J. 21(10–11), 1105–1121 (2000)

Fornell, C., Larcker, D.F.: Evaluating structural equation models with unobservable variables and measurement error. J. Mark. Res. **18**(1), 39–50 (1981)

Foss, K., Foss, N.J., Klein, P.G., Klein, S.K.: The entrepreneurial organization of heterogeneous capital. J. Manage. Stud. **44**(7), 1165–1186 (2007)

Foss, N.J.: The theory of the firm: the Austrians as precursors and critics of contemporary theory. Rev. Austrian Econ. **7**(1), 31–65 (1994)

Gassmann, O., Enkel, E., Chesbrough, H.: The future of open innovation. R&D Manag. **40**(3), 213–221 (2010)

Gebauer, H.: Exploring the contribution of management innovation to the evolution of dynamic capabilities. Ind. Mark. Manage. **40**(8), 1238–1250 (2011)

Gibbons, M., Limoges, C., Nowotny, H., Schwartsman, S., Scott, P., Trow, M.: The New Production of Knowledge: The Dynamics of Science and Research in Contemporary Societies. SAGE Publications (1994)

Gimeno, J.: Competition within and between networks: the contingent effect of competitive embeddedness on alliance formation. Acad. Manag. J. **47**(6), 820–842 (2004)

Gioia, D.A., Chittipeddi, K.: Sensemaking and sensegiving in strategic change initiation. Strateg. Manag. J. **12**(6), 433–448 (1991)

Granovetter, M.: Economic action and social structure: a theory of embeddedness. Am. J. Sociol. **91**(3), 481–510 (1985)

Granovetter, M.S.: The strength of weak ties. Am. J. Sociol. **78**(6), 1360–1380 (1973)

Gudergan, S.P., Ringle, C.M., Wende, S., Will, A.: Confirmatory tetrad analysis in PLS path modeling. J. Bus. Res. **61**(12), 1238–1249 (2008)

Guillén, M.F.: Models of Management: Work, Authority, and Organization in a Comparative Perspective. University of Chicago Press, Chicago (1994)

Gulati, R., Gargiulo, M.: Where do interorganizational networks come from? Am. J. Sociol. **104**(5), 1439–1458 (1999)
Hair, J.F., Sarstedt, M., Ringle, C.M., Mena, J.A.: An assessment of the use of partial least squares structural equation modeling in marketing research. J. Acad. Mark. Sci. **40**(3), 414–433 (2012)
Hair, J.F.J., Hult, G.T.M., Ringle, C.M., Sarstedt, M.: A Primer on Partial Least Squares Structural Equation Modeling (PLS-SEM). Sage, Thousand Oaks (2013)
Hamel, G.: The why, what, and how of management innovation. Harvard Bus. Rev. **84**(2), 72–84 (2006)
Hart, S.L., Dowell, G.: A natural-resource-based view of the firm: fifteen years after. J. Manag. **37**(5), 1464–1479 (2010)
Helfat, C., Finkelstein, S., Mitchell, W., Peteraf, M., Singh, H., Teece, D.J., Winter, S.G.: Dynamic Capabilities: Understanding Strategic Change in Organizations. Blackwell, Oxford, UK (2007)
Helfat, C.E.: Knowpabilities: understanding strategic change in mobility accumulation: the case of R&D. Strateg. Manag. J. **18**(5), 339–360 (1997)
Helfat, C.E., Peteraf, M.A.: The dynamic resource-based view: Capability lifecycles. Strateg. Manag. J. **24**(10), 997–1010 (2003)
Helfat, C.E., Winter, S.G.: Untangling dynamic and operational capabilities: strategy for the (n) ever-changing world. Strateg. Manag. J. **32**(11), 1243–1250 (2011)
Henderson, R.M., Clark, K.B.: Architectural innovation: the reconfiguration of existing product technologies and the failure of established firms. Adm. Sci. Q. **35**(1), 9–30 (1990)
Hertog, P.D., Aa, W.V.D., Jong, M.W.D.: Capabilities for managing service innovation: towards a conceptual framework. J. Serv. Manag. **21**(4), 490–514 (2010)
Ingram, P., Inman, C.: Institutions, intergroup competition, and the evolution of hotel populations around Niagara Falls. Adm. Sci. Q. **41**(4), 629–658 (1996)
Jansen, J.J.P., Bosch, F.A.J.V.D., Volberda, H.W.: Managing potential and realized absorptive capacity: how do organizational antecedents matter? Acad. Manag. J. **48**(6), 999–1015 (2005)
Keister, L.A.: Engineering growth: business group structure and firm performance in China's transition economy. Am. J. Sociol. **104**(2), 404–440 (1998)
Kim, L.: The dynamics of Samsung's technological learning in semiconductor. Calif. Manag. Rev. **39**(3), 86–100 (1997)
Kirzner, I.M.: Competition and Entrepreneurship. University of Chicago Press, Chicago, IL (1973)
Kogut, B., Zander, U.: Knowledge of The Firm, combinative capabilities, and the replication of technology. Organiz. Sci. **3**(3), 383–397 (1992)
Kohlbacher, M.: The impact of dynamic capabilities through continuous improvement on innovation: the role of business process orientation. Knowl. Manag. **20**(2), 71–76 (2013)
Kor, Y.Y., Mahoney, J.T.: Edith Penrose's (1959) contributions to the resource-based view of strategic management. J. Manage. Stud. **41**(1), 183–191 (2004)
Kor, Y.Y., Mahoney, J.T., Michael, S.C.: Resources, capabilities and entrepreneurial perceptions. J. Manage. Stud. **44**(7), 1187–1212 (2007)
Laursen, K., Salter, A.: Open for innovation: the role of openness in explaining innovation performance among U.K. manufacturing firms. Strateg. Manag. J. **27**(2), 131–150 (2006)
Lazonick, W., Prencipe, A.: Dynamic capabilities and sustained innovation: strategic control and financial commitment at Rolls-Royce plc. Ind. Corp. Change **14**(3), 501–542 (2005)
Lewin, A.Y., Massini, S., Peeters, C.: Microfoundations of internal and external absorptive capacity routines. Organ. Sci. **22**(1), 81–98 (2011)
Liao, J., Kickul, J.R., Ma, H.: Organizational dynamic capability and innovation: an empirical examination of internet firms. J. Small Bus. Manage. **47**(3), 263–286 (2009)
Lin, H.F., Su, J.Q.: Empirical research on the effect of dynamic capability on the process effectiveness of management innovation. Stud. Sci. Sci. **30**(12), 1900–1909 (2012)
Lin, H.F., Su, J.Q.: A case study on adoptive management innovation in China. J. Organiz. Change Manag. **27**(1), 83–114 (2014)
Lin, T.W.: Haier is higher. Strateg. Financ. (2009)
Lorenzoni, G., Lipparini, A.: The leveraging of interfirm relationships as a distinctive organizational capability: a longitudinal study. Strateg. Manag. J. **20**(4), 317–338 (1999)

Luo, J.D., Sun, Y., Xie, C.X., He, S.S.: Phenomena of capable persons in the functioning of self-organization. Soc. Sci. China **10**, 86–101 (2013)

McGrath, H., O'Toole, T.: Implementing a relational capability framework through an SME network. International marketing and purchasing conference Uppsala University (2008)

Mesquita, L.F., Anand, J., Brush, T.H.: Comparing The resource-base relational views: knowledge transfer and spillover in vertical alliances. Strateg. Manag. J. **29**(9), 913–941 (2008)

Mohr, J., Spekman, R.: Characteristics of partnership success: partnership attributes, communication behavior, and conflict resolution techniques. Strateg. Manag. J. **15**(2), 135–152 (1994)

Mol, M., Birkinshaw, J.: How management innovation happens. Mit Sloan Manag. Rev. **47**(4), 81–88 (2006)

Mol, M.J., Birkinshaw, J.: The sources of management innovation: when firms introduce new management practices. J. Bus. Res. **62**(12), 1269–1280 (2009)

Moorman, C., Zaltman, G., Deshpande, R.: Relationships between providers and users of market research: the dynamics of trust within and between organizations. J. Mark. Res. **24**(8), 314–328 (1992)

Nohria, N., Garcia-Pont, C.: Global strategic linkages and industry structure. Strateg. Manag. J. **12** (S1), 105–124 (1991)

O'Connor, G.C.: Major innovation as a dynamic capability: a systems approach. J. Prod. Innov. Manag. **25**(4), 313–330 (2008)

Penrose, E.: Theory of The Growth of The Firm. Wiley, New York (1959)

Podsakoff, P.M., Mackenzie, S.B., Lee, J.Y., Podsakoff, N.P.: Common method biases in behavioral research: a critical review of the literature and recommended remedies. J. Appl. Psychol. **88**(5), 879–903 (2003)

Powell, W.W.: Learning from collaboration: knowledge and networks in the biotechnology and pharmaceutical industries. Calif. Manag. Rev. **40**(3), 228–240 (1998)

Reimann, M., Schilke, O., Thomas, J.S.: Customer relationship management and firm performance: the mediating role of business strategy. J. Acad. Mark. Sci. **38**(3), 326–346 (2010)

Ridder, A.K.: Sensing and seizing open innovation: a capability-based approach. In: Paper Presented at the Dime-Druid Academy Winter Conference 2011 on Comwell Rebild Bakker. (2011)

Ringle, C.M., Wende, S., Will, A.: SmartPLS 2.0. University of Hamburg, Hamburg, Germany (2005)

Romijn, H.: Enterprise development in developing countries: innovation or acquisition of technological capability? In: Van Djik, M.P., Sandee, H. (eds.) Innovation and Small Enterprises in the Third World. Edward Elgar, Cheltenham (2002)

Rosenkopf, L., Nerkar, A.: Beyond local search: Boundary-spanning, exploration, and impact in the optical disk industry. Strateg. Manag. J. **22**(4), 287 (2001)

Rumelt, R.P.: Theory, strategy and entrepreneurship. In: D.J. Teece (ed.) The Competitive Challenge, pp. 137–158 (1987)

Salunke, S., Weerawardena, J., Mccoll-Kennedy, J.R.: Towards a model of dynamic capabilities in innovation-based competitive strategy: insights from project-oriented service firms. Ind. Mark. Manage. **40**(8), 1251–1263 (2011)

Sarstedt, M., Ringle, C.M., Smith, D., Reams, R., Hair, J.F.J.: Partial least squares structural equation modeling (PLS-SEM): a useful tool for family business researchers. J. Fam. Bus. Strateg. **5**(1), 105–115 (2014)

Schepker, D.J., Oh, W.-Y., Martynov, A., Poppo, L.: Themany futures of contracts: Moving beyond structure and safeguarding to coordination and adaptation. J. Manag. **40**(1), 193–225 (2014)

Schumpeter, J.A.: The Theory of Economic Development, vols. 1 and 2. Harvard University Press, Cambridge, MA (1934)

Singh, H., Zollo, M.: Globalization through acquisitions and alliances: an evolutionary perspective. In: Gatignon, H., Kimberly, J.R., Gunter, R.E. (eds.) The INSEAD-Wharton Alliance on globalizing strategies for building successful global business. Cambridge University Press, Cambridge (2004)

Souitaris, V.: Strategic influences of technological innovation in Greece. Br. J. Manag. **12**(2), 131–147 (2001)
Subramaniam, M., Youndt, M.A.: The influence of intellectual capital on the types of innovative capabilities. Acad. Manag. J. **48**(3), 450–463 (2005)
Szulanski, G.: Exploring internal stickness: to the transfer of best practice within the firm. Strateg. Manag. J. **17**(Special Issue), 27–43 (1996)
Tavani, S.N., Sharifi, H., Ismail, H.S.: A study of contingency relationships between supplier involvement, absorptive capacity and agile product innovation. Int. J. Oper. Prod. Manag. **34** (1), 65–92 (2014)
Teece, D.J.: Profiting from technological innovation: implications for integration, collaboration, licensing and public policy. Research Policy **15**(6), 285–305 (1986)
Teece, D.J.: Explicating dynamic capabilities: the nature and microfoundations of (sustainable) enterprise performance. Strateg. Manag. J. **28**(13), 1319–1350 (2007)
Teece, D.J., Pisano, G., Shuen, A.: Dynamic capabilities and strategic management. Strateg. Manag. J. **18**(7), 509–533 (1997)
Tiemessen, I., Lane, H.W., Crossan, M.M., Inkpen, A.C.: Knowledge management in international joint ventures. In: Beamish, P.W., Killing, J.P. (eds.) Cooperative Strategies: North American Perspectives, San Francisco (1997)
Tranfield, D., Starkey, K.: The Nature, Social Organization and Promotion of Management Research: Towards Policy. Br. J. Manag. **9**(4), 341–353 (1998)
Tu, Q., Vonderembse, M.A., Ragu-Nathan, T.S., Sharkey, T.W.: Absorptive capacity: enhancing the assimilation of time-based manufacturing practices. J. Operat. Manag. **24**(5), 692–710 (2006)
Vaccaro, I.G., Jansen, J.J.P., Bosch, F.A.J.V.D., Volberda, H.W.: Management innovation and leadership: the moderating role of organizational size. J. Manage. Stud. **49**(1), 28–51 (2012)
Van den Bosch, F.A.J., Van, W.R., Volberda, H.W.: Absorptive capacity: antecedents, models and outcomes. In: Social Science Research Network, ERIM Report Series Reference No. ERS-2003-035-STR (2003)
Volberda, H.W., Foss, N.J., Lyles, M.A.: Absorbing the concept of absorptive capacity: how to realize its potential in the organization field. Organ. Sci. **21**(4), 931–951 (2010)
Wang, C.L., Ahmed, P.K.: Dynamic capabilities: a review and research agenda. Int. J. Manag. Rev. **9**(1), 31–51 (2007)
Webster, F.E.: The changing role of marketing in the corporation. J. Mark. **56**(4), 1–17 (1992)
Wernerfelt, B.: A resource-based view of the firm. Strateg. Manag. J. **5**(2), 171–180 (1984)
Wiklund, J., Shepherd, D.: Knowledge-based resources, entrepreneurial orientation, and the performance of small and medium-sized businesses. Strateg. Manag. J. **24**(13), 1307–1314 (2003)
Williams, L., Rao, K.: Information technology adoption: using classical adoption models to predict AEI software implementation. J. Bus. Log. (1998)
Wilson, D.T.: Merging adoption process and organizational buying models. Adv. Consum. Res. **14** (1), 323–325 (1987)
Winter, S.G.: Understanding dynamic capabilities. Strateg. Manag. J. **24**(10), 991–995 (2003)
Witt, U.: Firms as realizations of entrepreneurial visions. J. Manage. Stud. **44**(7), 1125–1140 (2007)
Wu, Q., He, Q., Duan, Y., O'Regan, N.: Implementing dynamic capabilities for corporate strategic change toward sustainability. Strateg. Change. Special Issue: Entrepreneurship and Innovation **21**(5–6), 231–247 (2012)
Zahra, S.A., George, G.: Absorptive capacity: a review, reconceptualization, and extension. Acad. Manag. Rev. **27**(2), 185–203 (2002)
Zahra, S.A., Sapienza, H.J., Davidsson, P.: Entrepreneurship and dynamic capabilities: a review, model and research agenda. J. Manage. Stud. **43**(4), 917–955 (2006)
Zaltman, G., Duncun, R., Holbek, J.: Innovations and Organizations. Wiley, New York (1973)
Zollo, M., Winter, S.G.: Deliberate learning and the evolution of dynamic capabilities. Organ. Sci. **13**(3), 339–351 (2002)

Chapter 7
How Does Management Innovation Affect Performance of an Organization?

7.1 Introduction

Most scholars view the purpose of management innovation as an intention to further the goals of the organization, which may include both traditional aspects of performance and softer ones (e.g. Birkinshaw et al. 2008; Vaccaro et al. 2012). That is, the purpose of management innovation is to positively affect organizational goals and performance. However, the issue that how could intangible (adoptive) management innovation with ambiguous outcomes affect organizational performance still has not been addressed. The alignment theory between management innovation and tangible product or technology innovation indicates that this alignment has become a precondition for pursuing maximal performance of organizations (e.g. Daft 1978; Xu and Xie 2004), and management innovation may produce efficacy on organizations through product innovation. Management innovation serves the operation of an organization, while product innovation directly serves the market demand. This also indicates that tangible product innovation could produce direct effect on organizational performance, while management innovation may yield more continuous but lagging outcomes which offer strong support for product innovation.

Existing literatures have identified roles and advantages of both product innovation and management innovation. From a strategic perspective, innovation of new products well attuned to the voice of the customer, with perceived technical superiority, developed within budget and launched ahead of the competition, determines the market position of a firm (Calantone and Cooper 1979, 1981; Crawford 1980; Hultink et al. 1997); while introduction of new management practices is helpful in upgrading productivity of firms, improving the quality of customer offerings and creating long-lasting advantage for continuous development (Pil and MacDuffie 1996). However, in practice, the two types of innovation are not completely independent, but are always interlinked within the same complex organization system. In another word, an organization's innovation system consists

H. Lin, *Adoptive Management Innovation*,
https://doi.org/10.1007/978-981-10-7712-8_7

of both technological and administrative elements, and only when both elements are efficiently balanced could the value of innovation actually been achieved. For example, Vickery et al. (1999) examine the relationship between product customization (a key phase of product innovation) and organizational structure (a form of management innovation), and find that customization in product innovation associates with more formal control, fewer layers, and narrower spans of control. Researchers further identify the value of the alignment between technology or product innovation and management innovation (e.g. Daft 1978; Xu and Xie 2004). However, what is still puzzling is that how the alignment effect could be produced. Especially for those complex-product firms, how intangible management innovation with ambiguous outcomes could efficiently support the process of product innovation and finally further the goal of an organization is still a Black-box.

This chapter focuses on identifying major management innovation conducted in the innovation process of complex product with a separate brand and exploring how these management innovation support different phases of product innovation by considering two questions. First, what management innovation do in the process of product innovation more depend on, observable management practices or unobservable management ideas? Existing research focus either on observable management practices at the operational level (e.g. Birkinshaw et al. 2008; Mol and Birkinshaw 2009; Alänge et al. 1998) or on unobservable management ideas (e.g. Guillén 1994; Barley and Kunda 1992; Abrahamson 1996; Suddaby and Greenwood 2005). Some scholars refer to the former "because this is the level at which observable changes take place in the way work is done and the management innovation process can be witnessed" (Birkinshaw et al. 2008, p. 828); while others pay more attention to the latter because new management ideas would bring fundamental change for the firm. This research conducted a deep investigation into the whole innovation process of Lexus of Toyota Motor, aiming to find out all management innovation activities no matter observable or unobservable and analyze their functions. Second, how could management innovation efficiently support each phase of product innovation to produce alignment effect, and finally to improve the performance of an organization? The alignment mechanism between the two most important innovations in a firm has been developed as a significant research area focusing on elements integration of a firm in innovation. However, little has been down to explore how this alignment could be realized. In a word, this section aims to address how intangible management practices service the tangible process of product innovation.

This section is organized as follows. Part 7.2 describes the conception of product innovation, and presents a review of literatures on the alignment theory between management innovation and product innovation. Part 7.3 presents the method adopted including why the case study method adopted and how data is collected and analyzed. Part 7.4 shows the findings obtained from the case analysis on Lexus of Toyota Motor. Part 7.5 develops a framework of management innovation's support to the process of product innovation, and also discusses implications. Finally, Part 7.6 offers a brief conclusion.

7.2 Product Innovation and Management Innovation

7.2.1 Product Innovation

Since technology innovation (e.g. Henderson and Clark 1990; Utterback 1994) or tangible product innovation (e.g. Corsino and Gabriele 2011) is able to produce direct and explicit benefit (OECD 1991), the most research have focused on these innovations from both macro and micro analysis level of organizations. For example, Nayak (1991) illustrates the importance of product innovation for the increase of a company's profits, showing that the management of the product portfolio is of fundamental importance for the competitiveness of the firm. Evolutionary theories of economic change propose technology innovation and imitation a major driver of the relative performance of firms and the evolution of industrial structure (Nelson and Winter 1982). Corsino and Gabriele (2011) regard product innovation as a powerful factor behind differences in firms' performance, with companies that innovate successfully prospering at the expense of their less able competitors. Product innovation enables firms to better satisfy market demand and produce direct benefits, so it is ever more appreciated as a key component of sustainable growth for most firms in today's competitive marketplace (e.g. McNally et al. 2011; Cooper 2008).

A vast body of relevant researches have been produced which focuses on its process (e.g. Cooper et al. 1999, 2005; Cooper and Edgett 2008) and factors (e.g. Swink 2000; McNally et al. 2011). The most generic product innovation process is set up by Cooper and his research group in the early 1990s. They name the new product innovation process "the stage-gate process" (Cooper 1993, 2008), which is adopted and implemented in many U.S. industries during the late 1980s and early 1990s. According to Cooper (2008, p. 214), "A Stage-Gate process is a conceptual and operational map for moving new product projects from idea to launch and beyond-a blueprint for managing the new product development process to improve effectiveness and efficiency". The stage-gate process divides the whole process of product innovation into several sequential stages of discovery, scoping, build business case, development, testing and validation, launch, post-launch review, and so on. Similarly, Tzokas et al. (2004) set up a process including the generation of new product ideas, the development of an initial product concept, an assessment of its business attractiveness, the actual development of the product, testing it within the market, and the actual launch of the product in the marketplace.

Considering that new product innovation is risky due to alarming failure rates and the large amounts of venture capital required (Cooper and Edgett 2008), another group of scholars engage in identifying factors contributing to new product success. There are three widely accepted factors: time, quality, and expense (Bayus 1997; Smith and Reinertsen 1998; McNally et al. 2011). Speed to market is the time between idea generation and new product launch (Griffin 1993); Product quality refers to customer perceptions of superiority relative to competing alternatives (Sethi 2000); Development expense is the level of resources required for a project to advance from concept creation to commercial product (Clark and Fujimoto 1991). Meta-analyses suggest product quality is the most important factor of the three, followed by time considerations and R&D expenses (Henard and Szymanski 2001). McNally et al. (2011) expand the literature by accounting for the impact of cross-functional integration, and identify cross-functional integration as an element of integrated product development, a managerial approach to improve product innovation performance through overlap and interaction of product innovation activities. Additionally, factors referring to firm strategy, organizational structure and organizational changes also have been identified essential in new product innovation process (e.g. Henard and Szymanski 2001), which indicates that product innovation is a process relying on the whole organizational system. Though product innovation is a complex process, briefly speaking, it consists of three major phases of design and development, production and commercialization.

7.2.2 The Alignment Theory of Management Innovation and Product Innovation

There is no doubt that product innovation and management innovation are two distinctive groups of innovations (Birkinshaw et al. 2008). On one aspect, the process of management innovation is always ambiguous, compared with that of product innovation. They are relatively "difficult to observe, define and to identify system borders for" (Alänge et al. 1998, p. 8). Or as Mol and Birkinshaw (2006, p. 82) state, "in some cases it was impossible to say with any precision when management innovation actually took place". On another aspect, management innovation is typically tacit in nature and their results are always intangible, uncertain, lagging and even inseparable from that of technology innovation in organizations (Birkinshaw et al. 2008); while product innovation is typically explicit and their outputs are relatively tangible, certain and measurable.

Though these attributes suggest that management innovation and product innovation are quite different, the alignment between the two has been identified (e.g. Vickery et al. 1999; Daft 1978; Xu and Xie 2004). The concept of translation of ideas and the theory of fashion in innovation authored by Czarniawska and Sevon (1996, 2005) indicate that it is the organizational culture which contains mental and cognitive processes of the organizational personnel that imitate and support innovations. Or, organizational action is based on the apparatus of collective sense making and thus "change is a result of a blend of intentions, random events and institutional norms, as opposed to the idea of change as the result of strategic choice or environmental influence" (Czarniawska and Sevón 1996, p. 8). This cultural perspective of innovation, compared with evolutionary and teleological models which are not able to describe processes of innovation because of their being situated within the functionalist paradigm and thus operating on the assumption about natural adaptability to social (and market) environments, enables to describe processes connected with innovations pivotal for the contemporary organizations (Deetz et al. 2000). Accordingly, both management innovation and product innovation in an organization reflect the same organizational culture or mental and cognitive processes of the organizational personnel. This indicates a mutual base for the alignment between the two types of innovation.

More specifically, product innovation may stimulate the adoption and implementation of new management practices or ideas, and in turn, management innovation may promote product innovation (Calantone and Stanko 2007). When viewing management innovation and product innovation as two indispensable and efficiency approaches to gain continuous competitive advantages, a firm needs to keep balance between the two in resource allocation and strategy setting. For example, throughout the new product innovation process, managers need to master managerial techniques for planning, development, deployment, evaluation and control to increase the success rate of their new product efforts. To do so, they need to align their new product strategy with their corporate strategy and secure the focusing of their product innovation processes to the strategic imperatives of the firm. Daft (1978) sets up a duel-core model of innovation which postulates that an administrative core and a technical core coexisting in the organization. Though the administrative core sets the functioning of the structure and coordination activities, and the technical core serves to transform inputs, they act as integrated subsystems to improve organizational performance. Miles and Snow (1978) argue that the strategy adopted by an organization is a powerful determinant of its willingness and capacity to develop new products. According to the authors, organizations must solve their adaptive cycle by congruently aligning two internal problems: the administrative problem and the technological or product problem. This alignment may enable a firm to more efficiently innovate. Vickery et al. (1999) examine the relationship between product customization (a key aspect of product innovation) and organizational structure (a form of management innovation), and find that customization in product innovation associates with more formal control, fewer layers, and narrower spans of control. Similarly, Xu and Xie (2004) further the alignment between product innovation and management innovation by adopting a case study on the twenty-year development process of Hair (A Chinese appliance provider, also the No. 1 global home appliance brand), and they find that the fast development of Haier depends on disruptive innovation in both its product solutions and management model. For example, Haier has changed its organizational structure from a pyramid into an inverted pyramid and then flattened it even further to become a dynamic network-based organization composed of innovative ZZJYTs (self-management unit), in order to support product innovation in different periods of time. Following these studies, some other scholars have further explored the value of this alignment, and confirmed it a precondition for achieving maximal performance of organizations (e.g. Tidd et al. 2001).

However, the literature discussing the alignment between management innovation and product innovation contains gaps. Though both product innovation and management innovation are of importance and they are interconnected and interdependent, what is still puzzling is that how this alignment could be achieved.

According to the concept of alignment, it consists of two key questions: how product innovation enables to stimulate changes in management practices or ideas; and how management innovation enables to support product innovation. Here, in this research, the author attempts to address the latter one. In particular, for those firms offering complex products and dominated by product innovation, how could intangible management innovation efficiently supports the whole process of product innovation? To do so, a case study on Lexus of Toyota Motor was conducted in this chapter. To be noticed, since both generating and adoptive management innovation are able to produce similar effects on organizations, this chapter refers to both two types of innovation, as in Chap. 4.

7.3 Method

7.3.1 The Case Study

As discussed in Sect. 7.3, an exploratory single case study was adopted to exert its function of theories exploration and construction (Eisenhardt 1989; Yin 1994; Eisenhardt and Graebner 2007). Since not only gaps do exist in the literature of management innovation but also little research goes into exploring how management innovation support product innovation, an exploratory case study might be appropriate for developing a theoretical framework from case-based and empirical evidence.

7.3.2 The Case Sample

The research setting was Toyota Motor Corporation Limited (Toyota Motor), which had grown from a niche player in the cheap compact cars into the world's the largest automobile manufacturer since 1980s. During the financial year of 2013[1] (April 2012–March 2013), it made sales of 22,064.1 billion yen, operating income of 1320.8 billion yen, and net income of 962.1 billion yen by producing over 8.6 million and selling over 8.8 million models in both Japan and overseas. Since its founding in 1937 by Toyoda Kiichiro, Toyota Motor had sought to contribute to a more prosperous society through the manufacture of automobiles, operating its business with a focus on vehicle production and sales. As an innovative leader, Toyota Motor was well-known for both products and management philosophies. A series of new products have been launched to the market every year by Toyota Motor, for example, Vanguard, Mark X ZiO, Highlander, Highlander Hybrid and Scion xD in 2007, Crown Hybrid, Vellfire and iQ in 2008, and Aqua in 2011.

[1]Data sources: http://www.toyota-global.com/company/profile/overview/.

Meanwhile, Toyota Motor always worked hard to keep its advantage in management. For example, the Toyota Production System, also referred to as a "lean manufacturing" or "just-in-time" system, had come to be respected and studied worldwide; The Creative Idea Suggestion System launched in 1951, with the number of suggestions made steadily increased, had supported flexible responses to changes that involve *monozukuri* (conscientious manufacturing); the basic concepts of TQM and problem solving as well as *kaizen* (continuous improvement) through creative innovation had spread throughout the company and taken root, contributing to higher product quality and work quality at all levels and improving the vitality of individuals and organizations. Hamel (2006) even argues that the answer for why it has taken America's automobile manufacturers so long to narrow their efficiency gap with Toyota is the radical management principle at the heart of Toyota's capacity for relentless improvement. In a word, the success of Toyota Motor lied on both product (technology) innovation and management innovation.

Toyota Motor owned two world-popular vehicle brands, Toyota brand and Lexus brand. The rapid development of Toyota Motor and its successful expansion to World's luxury car market might be attributed to the efforts made to Lexus brand. First introduced in 1989 in the United States, Lexus had been sold globally over 70 countries and territories and become Japan's largest-selling make of premium cars. Though Lexus exceeded and became America's best-selling line of luxury motor vehicles barely a dozen years after its launch, it had become a global luxury model after an organizational separation from parent company Toyota Motor in early 2000s. When Lexus faced an identifying crisis, with dedicated design, engineering, training, and manufacturing centers working exclusively for the division, the purpose of this reorganization was to better synchronize the activities of recreating Lexus, from designing to commercialization. Hence, Lexus vehicles had been redesigned as "global models" for international release. Therefore, this reorganization indicated not only a series of management innovation, but a process of product innovation.

In brief, the case of Toyota Motor's Lexus was adopted for two reasons. One is that, Toyota, which had dedicated in manufacturing low-and-medium-scale vehicles for over 50 years before 1980s, has successfully launched the best-selling luxury brand of Lexus within 30 years in the upscale automobile market which had been dominated by well-known brands of Benz, BMW, Audi and Porsche. This indicates an unprecedented challenge and a big success. The other is that, when going through the whole process of Lexus redefinition since 2001, the author finds that the success of Lexus depends not only on its high capability of technology innovation, but also that of management innovation. Therefore, this research focuses on the whole recreation process of Lexus, i.e., a process which promotes expansion of Lexus to the world luxury car market, to explore how Toyota Motor conducted various observable and unobservable management innovation to promote each phase of the whole innovation process of complex product.

7.3.3 Data Collection

The data collection lasted from March 2009 to May 2013. Three primary data collection mechanisms (archival data, semi-structured interviews, and observation) were used to meet criteria for trustworthiness (Lincoln and Guba 1985; Yin 1994) and help create a rich understanding of how Lexus was launched and how management innovation serviced the process of product innovation. Moreover, a combination of primary data with secondary data was made in data collection for this research. Primary data and secondary data are two sets of data collections for research identified by Hox and Boeije (2005). Primary data are original data for a specific research goal; while secondary data represent data originally collected for a different purpose and reused for another research question (Hox and Boeije 2005). Using primary data enables researchers to collect their own data information for the specific purposes of their study, so that "the operationalization of the theoretical constructs, the research design, and data collection strategy can be tailored to the research question, which further ensures that the study is coherent and that the information collected indeed helps to resolve the problem" (Hox and Boeije 2005, p. 594). For this reason, a large number of researchers advocate primary data, particularly in case studies. However, primary data are always costly and time-consuming. So, if relevant information on a new research topic is accessible, reusing it would be a good choice. That is, secondary data are appropriate when they are able to answer the newly formulated research question, or provide the researcher with a wider sample base for testing interpretation at far less cost and with greater speed (Yin 1994; Hox and Boeije 2005). Considering that to collect primary data on innovation of Lexus from Toyota Mobile particularly from its headquarters in Japan was too costly and difficult, and existing information on Lexus and its innovation process was accessible and sufficient to answer the research question, the author adopted secondary data collected through archival data as the major source for data in this research. In order to add more information to the research, the author also conducted some interviews on employees from Toyota Mobile China and Dalian Zhongsheng Lexus Automobile Sales and Service, and took notes of observations.

Archival data In total, Eight-two documents representing 652 pages of data in forms of the innovation proposal, innovation notes, newsletters, memos, annual reports and innovation performance reports were collected. Most of these data were collected through international magazines and articles, writings,[2] media reports,

[2]These writings particularly included *Lexus Miracle* written by Takagi Haruo (2010), and *LEXUS: The Relentless Pursuit (Revised Ed.)* by Chester Dawson (2011). The former focuses on how Lexus was redefined and launched to the world market; and the latter focuses on describing what drove the Lexus brand's success (i.e., the how's and why's of its reputation for top-notch quality, the unforgettable advertising campaigns and bespoke customer service), and also the inside story on the growing pains of Toyota Motor's luxury division, its vehicle development plans and the lawsuits that almost derailed the brand.

stories and web materials related to Toyota and Lexus. Also, the Institution Office of Toyota Mobile China gave the author access to some information on its management and product innovation. The author created categories for filing, retrieving, and analyzing the archival data, separating product innovation and management innovation documents. These data not only offer the general information about Toyota Motor and Lexus (for example, company profile, global vision, management philosophies, production system, the relationship between branches and the headquarters in Japan, its marketing and product strategy, the balance between Toyota brand and Lexus brand), but also enable the author to track innovation process of Lexus, major management innovation and alignment of various management innovation with the process.

Semi-structured interviews Two sets of interviews were conducted and transcribed. The first set with managers and employees from Toyota Mobile China Investment (i.e., a training program in Dalian University of Technology), who knew much about Toyota Motor and Lexus, and more importantly, could be easily contacted, was conducted during March 2011–September 2011. This set of interviews focused on technology advantages of Toyota and the technology difference between middle-and-low scale Toyota brand and upscale Lexus brand, and also offered a rich understanding of how Lexus entered the market of China and other areas and what technology advantages of Lexus over vehicles of Toyota brand and other upscale vehicles. The second set of interviews occurred during November, 2011–May, 2012 with general managers and employees of Dalian Zhongsheng Lexus Automobile Sales and Service in China, aiming to get information on specialties of Lexus sales and service and the difference from Toyota brand. This set of interviews consisted of one face-to-face interview and one telephonic interview with the general manager of Dalian Zhongsheng Lexus Automobile Sales and Service, another two face-to-face interviews and one group interview with its employees. These interviews ranged 60–90 min in length. Interviews, especially these with managers, began with questions covering more general topics. In view of the inductive aims, the author encouraged them to go deeply into details. Before each set of interviews, an interview protocol was designed with major themes in mind; and during the interviews, questions were not asked in any specific order but were governed instead by the actual situation (Gummesson 2000).

Observation During each visit, the author made notes of the informal observations while waiting for interviews and walking around. The sales and service store, where various Lexus models were exhibited and customers were offered with service directly, offered the opportunity to observe how marketing strategies serviced commercialization of Lexus vehicles.

7.3.4 Data Analysis and Coding

The author and another team member participated in data analyzing and coding. The process of analysis was guided by Eisenhardt's notion that "it is the intimate connection with empirical reality that permits the development of a testable, relevant, valid theory" (1989, p. 532). It proceeded in six stages. Firstly, after being collected, the data was sorted, with rich raw data transformed into the written form. Archival materials offered abundant information on the case and expanded the understandings of the case by offering insights that might refute or reinforce the findings (Andriopoulos and Lewis 2009). Secondly, via blending all data on Lexus product innovation together and taking it into a sequential order, a story of Lexus innovation was written. The intention was to capture the ebb and flow of important activities, identify key activities in each phase and set up a process framework of Lexus product innovation from design and development, to production and commercialization. In the third stage of analysis, explicit management innovation (i.e., observable management practices) were identified, also through iterative comparisons and coding, including setting up of Business Reform Institution, Lexus Center (R&D), Liaison Councils, Lexus Production Center and Lexus Marketing Department. Fourthly, implicit management innovation (i.e., unobservable management ideas) were identified, including product strategy innovation of Business Reform Institution, design and development conception innovation of Lexus Center, improved lean production of Lexus Production Center, marketing strategy innovation of Lexus Marketing Department, and horizontal communication mechanism innovation of Lexus Councils. Fifthly, the author and the team member engaged in exploring how these explicit and implicit management innovation practices promoted activities of product innovation in each phase and made Lexus a big success. Finally, a model showing the alignment of various management innovation with complex product innovation process was set up.

In this process of analysis and coding, two techniques were used to address the problem that "no analysis strategy will produce theory without an uncodifiable creative leap, however small" (Langley 1999, p. 691). The one is intervention of other two research coworkers to ask critical questions and introduce alternative explanations of the data to improve the quality of the theorizing. As an outsider to the site, they identified patterns in the data that the author either supported or refuted by using their rich understanding of the data. The second is analysis techniques such as constant comparison (Glaser and Strauss 1967) and content analysis (Krippendorff 2004). These techniques were used not only to enhance the interpretation of the data, but also to increase the confidence in the analytical process (Golden-Biddle and Locke 2007). For example, in the third stage, the author and the member identified major management innovation activities separately and then compared the coding results with each other to reach a consensus.

Systematic, iterative comparisons of data, emerging categories, and existing literature aided development of cohesive constructs and an integrative, theoretical framework of the support of management innovation to the whole process of

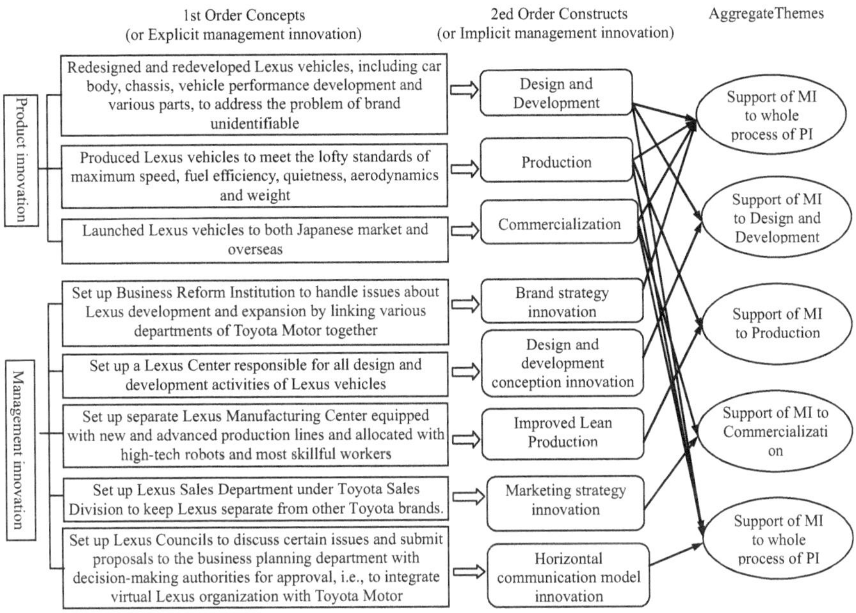

Fig. 7.1 Data structure. *Note* MI represents management innovation, and PI represents product innovation

product innovation. As for product innovation, the codes discerned similar were collated into first-order activities, and then linkages among the activities that could lead to the development of second-order constructs were discerned by formulating researcher-induced concepts at a more abstract level. As for management innovation, observable structure changes were collated into explicit management innovation as first-order activities, and then unobservable management ideas produced by these set-up structures were explored and identified as second-order constructs. Finally, the second-order constructs of product innovation and implicit management innovation were connected to show the internal alignment mechanism and a framework explicating the relationships was developed.

Figure 7.1 shows the data structure of the findings. It depicts the five main aligning points of management innovation's support to product innovation. More specifically, it presents three phases of product innovation (2nd Order Constructs) and their first-order concepts (for example, Redesigned and redeveloped Lexus vehicles, including car body, chassis, vehicle performance development and various parts, to address the problem of brand unidentifiable; produced Lexus vehicles to meet the lofty standards of maximum speed, fuel efficiency, quietness, aerodynamics and weight), explicit management innovation (for example, set up Business Reform Institution to handle issues about Lexus development and expansion by linking various departments of the Toyota Motor organization together; set up Lexus Center responsible for all design and development activities of Lexus vehicles) and implicit management innovation (for example, brand strategy innovation, design and development conception innovation), as well as aggregate themes of alignments (for example, support of management innovation to whole process of product innovation, support of management innovation to design and development).

7.4 Findings

This section will discuss the findings in detail by exploring deeply into each phase of product innovation and how both explicit and implicit management innovation support each phase.

7.4.1 Business Reform Institution, Brand Strategy Innovation and Product Innovation

Toyota Motor was keenly aware of the potential for stagnation at Lexus and undertook efforts to give the brand a clearer identity in the marketplace and some much-needed vigor in the design department. To do so, an explicit innovation of setting up Business Reform Institution was done in the early time of reorganization by linking various departments of the Toyota Motor together. Though it was a temporal unit operating as a project team, it held more responsibilities and authorities. Business Reform Institution belonged to the division, consisting of offices and lower-level groups. In order to keep Business Reform Institution a separate unit, Toyota Motor appointed employees from different departments to concurrently occupy the positions. Therefore, Business Reform Institution was not subordinate to any department. It was an effective form of structure flattening and cross-functional integration.

Within this new observable unit, an unobservable brand strategy was proposed. As a unit reporting directly to top managers of Toyota Motor, Business Reform Institution was responsible for all long-term planning activities related to Lexus brand strategy innovation and proofing. The primary task was to reposition Lexus as a global luxurious upscale car brand rooted in Japanese Culture, and reset up the goal that it was not just to match but to exceed the lofty standards set by Benz and BMW in five key areas: maximum speed, fuel efficiency, quietness, aerodynamics and weight. Business Reform Institution suggested a separation of Lexus from its parent company Toyota Motor to keep a single upscale image from that of ordinary vehicles with lower prices, with the purpose of preventing negative effect of low-and-medium-scale image of Toyota brand on the new luxurious and upscale Lexus brand.

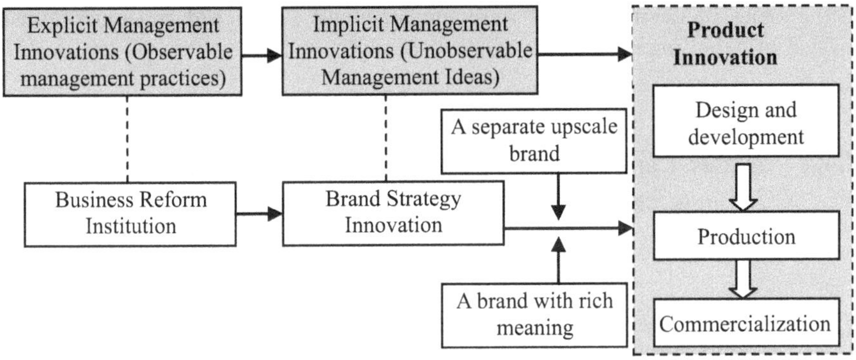

Fig. 7.2 Business Reform Institution, brand strategy innovation and product innovation

> Though the etymology of the Lexus name had been attributed to the combination of the words 'luxury' and 'elegance', and it was an acronym for 'luxury exports to the U.S.', the brand name had no specific meaning and simply denotes a luxurious and technological image. (LEXUS: The Relentless Pursuit)

Business Reform Institution further explored the nature of upscale cars, offered Lexus brand with rich meaning and proposed effective means of pursuing luxury, for example, treating continuous offering high-quality vehicles with best sales and service as a fundamental strategy for pursuing luxury; proposing four specific means of stressing the feeling of luxury, serving every customer with heart and soul, making coexisting of two opposite laws (for example, pursuit of both maximum speed and quietness) and offering best products to pursue luxury; making a commitment to make the most of every moment with Lexus; seeing every employee representative of Lexus, working with creation, confidence and enthusiasm. In a word, the new brand strategy of Lexus made by Business Reform Institution focused on separate upscale brand positioning and rich brand meaning offering.

This brand strategy innovation might affect the whole process of product innovation, as shown in Fig. 7.2. Following the direction of brand separation and rich luxury meaning, new structures of Lexus Center, Lexus Manufacturing Center, Lexus Marketing Department and Lexus Councils were also reset up separating from Toyota Motor. In the process of design and development, great attention was given to every detail of car body design, comfort, driving dynamics, safety and operation, to meet the new brand standard of pursuing luxury, serving every customer with heart and soul, and making customers full of passion and movement; in the process of production, the separate Lexus Manufacturing Center and lines focused on each detail and component and part by adopting higher-quality controlling system and operation process to stick to standards of World-class luxury cars; similarly, when launching new Lexus vehicles, Toyota Motor set up a new marketing strategy focusing on making customers feel luxury and perfection, movement and passion through a separate marketing department. In addition, all technologies of Lexus vehicles were fundamentally redesigned and redeveloped, instead of being improved from existing technologies.

7.4.2 Lexus Center, Design and Development Conception Innovation and Product Innovation

> If there is one area where I and our owners see a major opportunity to take Lexus to a new level in luxury, it is in styling. (The vice president of Lexus Division)

In order to redesign and redevelop new Lexus vehicles, an observable management practice of setting up Lexus Center was conducted. Lexus Center, responsible for all design and development activities, was subordinate to R&D division of Toyota Motor, parallel to Business Planning Department of Toyota brand. Before the reorganization, R&D division of Toyota Motor was made up by three centers: the first center was responsible for rear wheel driving cars and sports cars; the second center was responsible for front wheel driving cars; and the third center was for commercial vehicles and SUV (Sport Utility Vehicle). As more and more models of Lexus vehicles developed, Lexus models were scattered across different centers, for example, LS in the first center, ES in the second center and LX in the third center, leading to the disintegration of Lexus design styling and disunity of Lexus brand. Therefore, Lexus Center, consisting of Lexus Business Planning Department, Single Model Developing Teams (for example, LS, GS, ES, IS, IS C, CT, HS, RX, GX, LX, LFA, IF S), Car Body Design Department, Chassis Design Department, and Vehicle Performance Development Department, was set up to keep styling of Lexus vehicles away from that of other Toyota vehicles. Business Planning Department, serving as a critical strategic point in Lexus division, though was set in Lexus Center, it focused not only on technology research and development, but more importantly on planning of Lexus brand. Members from departments of design, customer service and production were selected to joint this department. A Lexus brand planning office was also set up under this Business Planning Department to keep Lexus separate, upscale and special. Though Lexus Center and Business Planning Department of Toyota brand operated independently, they always coordinated to address certain technological issues.

A series of implicit management innovation referring to design and development philosophy were conducted within Lexus Center. In order to provide more gravitas based on Japanese strengths in design and to counter criticism that Lexus vehicles were too derivative of other makes and models, Lexus Business Planning Department promoted a new styling language termed "J-factors" design gospel which represented a blend of the old and the new, with samurai aesthetics mixed in equal parts with anime quirkiness and video-game high tech. It meant incorporating what was known as "L-Finesse", an oblique concept conjured up to give designers more styling direction without short-circuiting their creative impulses. L-Finesse incorporated three core themes:

> Seamless anticipation. Anticipating the customer's needs and desires is central to Lexus, and to the Japanese spirit of hospitality. By integrating anticipation into design, forms that go beyond three dimensions are created, resulting in unique and personal experiences for all our customers.
>
> Incisive simplicity. The simplicity of Lexus lies not in mere reduction of form. The removal of extraneous elements reveals the purity and beauty that is condensed within the structure and surface of the vehicle.
>
> Intriguing elegance. Lexus elegance reveals itself through careful juxtaposition of opposing elements. The result is an experience filled with fascinating depth, which appeals to the senses through a profound and mysterious beauty. (www.lexus-int.com)

The purpose of setting up Single Model Developing Teams was to focus on building special image strategy for each model, so as to further guide design and development activities. For example, the IS (Intelligent Sport) team defined IS as an entry-level sport model which represented sensibility and coexisting of impressive aerodynamics, quietness and pleasure. So this compact vehicle was produced using a shortened, front-engine, rear-wheel drive midsize platform, allowing Japanese buyers to take advantage of tax savings imposed by Japanese government regulations concerning vehicle exterior dimensions and engine displacement. LS, whose image strategy was developed by the LS team, was defined as a full-size luxury sedan that served as the flagship model of Lexus. All LS vehicles were equipped with V8 engines and rear-wheel drive, and since 2006 all-wheel drive, hybrid, and long-wheelbase variants had been offered. Each model had its own image strategy established by certain model developing team.

Finally, Car Body Design Department, Chassis Design Department, and Vehicle Performance Development Department were set up to pursue luxury by focusing on every technological component of a sedan. Each department was allocated with specialized personnel. These departments proposed a philosophy of continuous improvement by "making coexisting of two opposite laws" in designing. For example, Car Body Design Department, which was responsible for both exterior and interior design, attempted not only to upgrade existing designs, for example, upgrading materials (such as a headliner made out of ecsaine, a suede-like material used by couturier Yves Saint-Laurent), reducing the forward and aft coefficient of lift, but also created new designs, for example, adaptive headlamps housed in hand polished surrounds designed to look like Baccarat crystal tumblers, chrome exhaust vents integrated into the rear bumper in LS 460, a 24-h concierge/emergency aid service and Lexus Link offered in North America, with the analogous G-Link system offered in Japan.

> We have 30 main choices of Lexus body colors, but the LFA (a two-seat supercar from Lexus) has unlimited options. Even so, around half of LFA customers choose white. Black and silver are very popular Lexus choices, so more effort is going into enriching and differentiating these colors with unique textures and depths. (A color designer)

By using carbon fiber woven in-house and reinforced with plastic resin for the frame, Toyota Motor was able to make the cars 221 lb lighters than an equivalent aluminum chassis. The new cutting-edge body technology was used, which lent itself to the LFA's exterior styling flourishes like cornered vents and the almost razor sharp creased, and helped lower the vehicles' Cd to an impressive 0.31. This indicated that each body technology needed continuous improvement and innovation. Similarly, as for chassis design, the specialized department engaged in improving four core systems of transmission, driveshaft, operation and braking. For example, the Intelligent Parking Assist System feature was equipped in LS 460 which can parallel-park or reverse-park the LS into a preselected space with minimal brake input at the push of a button. Dynamic Radar Cruise Control could accelerate and brake while monitoring traffic, a Brake Hold button, which prevented creeping forward motion when the driver's foot was off the brake pedal, and the Automatic Parking Brake, which could engage the parking brake simultaneously whenever the transmission is shifted to Park. Vehicle Performance Development Department focused on improving the full vehicle performance, module performance, subsystem performance and component performance. In a word, it engaged in performance activities of general layout, ergonomic, styling and color, power performance, fuel economy, steering, and so on.

As shown in Fig. 7.3, Lexus Center, an explicit management innovation, affected the design and development phase of product innovation through a series of tacit management innovation. Lexus Business Planning Department promoted the

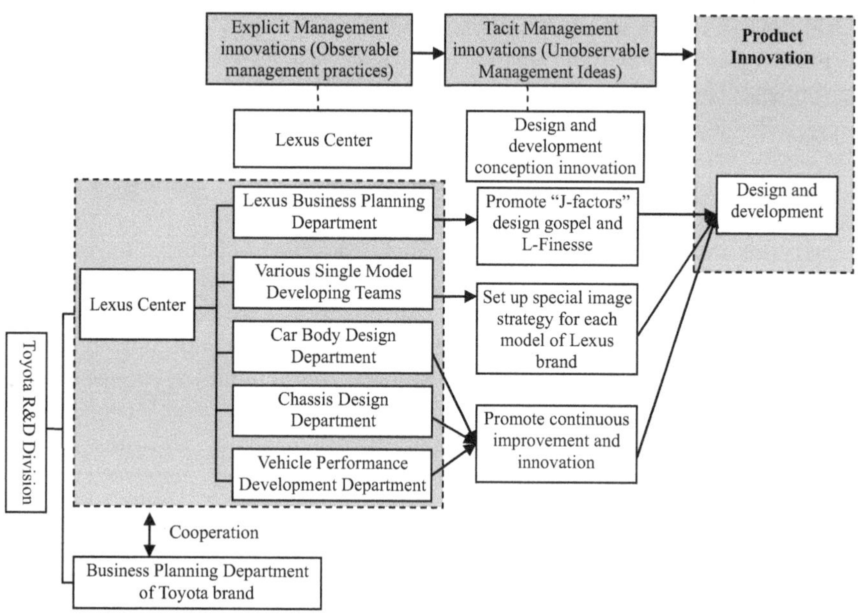

Fig. 7.3 Lexus Center, design and development conception innovation and product innovation

"J-Factor" design gospel and "L-Finesse" design philosophy; Single Model Developing Teams created special image strategies for each model of Lexus; and the rest departments created designing philosophy of "making coexisting of two opposite laws". All these unobservable ideas could promote new product design and development.

7.4.3 Lexus Manufacturing Center, Production Philosophy Innovation and Product Innovation

Catering for the high-quality demand of Lexus vehicle, Lexus Manufacturing Center was set up to produce new Lexus products. The new center was equipped with new and advanced production lines, and allocated with high-tech robots and most skillful workers. Various unobservable production philosophies were promoted by this explicit practice to promote product innovation.

For one aspect, a philosophy of further Lean Manufacturing was introduced into Lexus division. Lean Manufacturing, a production practice that considered the expenditure of resources for any goal other than the creation of value for the end customer to be wasteful, and thus a target for elimination, allowed Lexus division to reduce its parts inventory and efficiently produce only precise quantities of items. With its emphasis on continuous improvement, the value of employee commitment and superior quality, Leaning Manufacturing had been seen as a key element for the success of Lexus. The approach to Lean Manufacturing focused on improving the "flow" or smoothness of work, thereby steadily eliminating mura ("unevenness") through the system and not upon "waste reduction" per se. Specific principles of Lean Manufacturing implemented in Lexus Division included Pull processing, Perfect first-time quality, Waste minimization, Continuous improvement, Flexibility, Autonomation, Load leveling and Production flow and Visual control. In comparison with the Lean Manufacturing in Toyota Motor, the improved philosophy focused more on quality and continuous improvement through a whole process from designing a simple manufacturing system, to recognizing the room for improvement, and to continuously improving the manufacturing system design.

For another aspect, a new manufacturing concept of Lexus Production Engineering Advanced Craftsmanship Effort (L-PEACE) was promoted to integrate advanced production technologies with superior craftsmanship. Lexus vehicles were created with a masterful skill and attention to details inspired as much by traditional artisanship as by state-of-the-art engineering. This pursuit of perfection was evident at every stage of the manufacturing process. For example, at the Tahara, Japan facility, components precision-made with leading edge digital technology were assembled and then finished by master craftsmen called Takumi. A Takumi had a synergy of technical expertise and sure senses honed by decades of rigorous experience. At the vanguard of quality, every Lexus was a masterpiece of automotive artistry brought to life with human hands sensitive to the heartbeat of

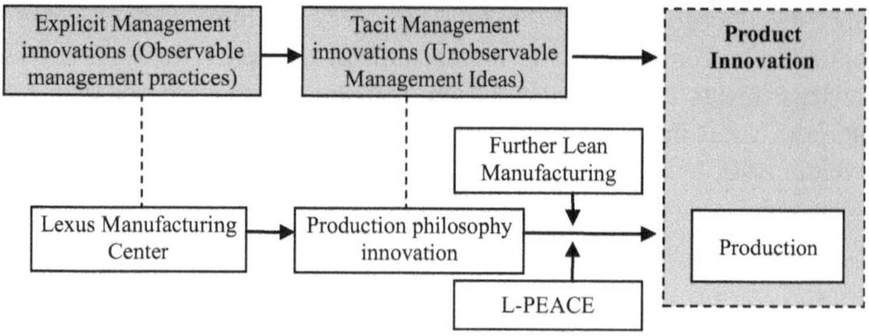

Fig. 7.4 Lexus Manufacturing Center, production philosophy innovation and product innovation

innovative technology. Master craftsmen ensured that each individual part of the total, mind boggling 30,000 that make up every vehicle was bolted together as precisely as the original blueprints, strictly checked, slightly adjusted and meticulously inspected. And not only did they keep the Lexus tradition alive by teaching skills, processes, and the industry's highest quality standards—they were also constantly learning, with an eye out for new ways to nudge their creations to even greater heights. Take Sakakibara, Lexus's paint guru, for example, he was able to spot the tiniest of blemishes in a paint job that other carmakers would otherwise overlook. Though robots sprayed up to six coats of paint on the body of every car, Sakakibara knew that robots were only as good as the specialists who did the programming. The more experienced and knowledgeable the takumi behind the machine, the more nuanced and sensitive the machine's work.

> We cannot ever let our vigilance lapse. The people who buy our cars have sharp eyes and an appreciation of beautiful lasting quality. (Sakakibara, Lexus's paint guru)

As shown in Fig. 7.4, Lexus Manufacturing Center, an explicit management innovation, affected the production phase of product innovation through an implicit philosophy innovation of further Lean Manufacturing and L-PEACE.

7.4.4 Lexus Sales Department, Marketing Strategy Innovation and Product Innovation

Lexus Sales Department, an explicit structure change, was set up under Toyota Sales Division to keep sales of Lexus separate away from that of Toyota brand. Best management resources were allocated to this department, for example, appointing Yokoi Yasuhiko who had been responsible for Lexus marketing in the European market since 1990s as its head, Maruyama who had participated in Lexus project for more than ten years as the group leader, and Kimura Takano who had 14-year oversea marketing experience to undertake the task of employee trainings. Treating commercialization as a key phase for Lexus, Lexus Sales Department conducted a series of implicit innovations.

The general marketing strategy Lexus Sales Department attempted to transform the existing huge-crowd marketing tactic and local war tactic to uniform brand strategy. The traditional Toyota sales system was of geographical layout in Japan, where distributors in the same region carried out various marketing activities independently. As a result, rivals of a certain distributor referred to not only distributors of other automobile manufacturers, but also other distributors of Toyota brand. The headquarter of Toyota Motor allowed the existence of this local or internal competition. Additionally, in the traditional pattern, Toyota Motor tried to cover the whole market through continuously setting up new sales stores in potential-customer-concentrated areas, and to employ as many salesmen as possible to offer a better approach. In contrast, the marketing strategy of Lexus had been transformed to be more quality-dependent instead of quantity-dependent. Lexus Sales Department focused on the group of potential customers with income of 200,000–300,000 instead of the whole market. A total of only 143 Lexus sales stores had been set up till August 2005 when Lexus was launched into Japanese market. Until now, the number has not exceeded 200.

The relationship with agents In the traditional marketing pattern of Toyota Motor, sales stores did not get directions from Toyota Sales Department but through the medium of regional agents. Regional agents tried to understand the intention from Toyota sales department about layout of stores, opening style, customer information and demands, and then delivered it to stores. In contrast, in the new pattern of Lexus, sales stores were under the control of Lexus Sales Department directly to make sure the consistence of Lexus brand. Lexus Sales Department was in charge of establishing selling conception, measures and approaches, conducting order management and employees recruiting and training, instead of regional agents. This new relationship helped to shorten the distance between the firm and its customers. In the traditional pattern, regional agents who owned abundant of sales stores tended to set up differentiated selling strategies; while in the new pattern, all sales stores of Lexus tended to hold the same selling conception and strategy, and offer standard services.

Employee selection and training A series of new measures were adopted by Lexus Sales Department in employee selection and training. First, the storekeeper of Lexus was retitled as the general manager, with their roles transformed from executives to managers. Hence, the general manager of a Lexus sales store was not only a business runner, but also a Lexus brand spokesman and a talent cultivator. They were also responsible for proposing standards for sales managers, sales consultants, workshop supervisors, technicians, receptionists and assistants.

Second, three sets of standards (i.e., expertise, competency and responsibility) were proposed for selecting the best general managers. Expertise referred to particular automobile knowledge and skills held by candidates; competency indicated high-quality services offered by candidates; responsibility stressed capability of "making customers the most of every moment" in selling through charming smiling, willingness to communicate with customers and caring for customers. The three sets of standards were also suitable for selecting other employees, with relevantly lower requirements. Third, a series of training programs were offered. For example, the training program for general managers referred to: a Lexus brand theory training; a U.S research journey to experience Lexus selling atmosphere; a counter service training in The Ritz Carlton (Osaka) Hotel and senior department store to learn how to offer high-quality services; a product training during which the manager of Lexus Center, R&D team representatives and designers would present the history of Lexus, characteristics of Lexus vehicles and their technologies and designing philosophy to general managers; an opportunity to visit Lexus manufacturing workshops; and a series of business college courses. As for other employees, besides brand conception learning, test drive, and workshop visit, the training program focused on service offering and responsibilities of employees, such as phone answering skills, marketing process, greeting etiquette and negotiation skills.

Store layout and service All Lexus stores were located in the most prosperous areas of cities. Lexus Sales Department attached importance not only to the exterior and exhibition room, but also to subsidiaries like customer lounge. A glass wall was used for exhibition room to make sure that customers could catch a sight of Lexus vehicle from a distance. The service counter was set up in the front of exhibition room to offer immediate service to customers. More importantly, special reception rooms were set up in the best location to offer customers with private room and make them relax and pleasure. In a word, the layout of Lexus store embodied the concept of respecting privacy of customers and offering comfortable room for customers. In any Lexus store, customers could get standard and considerate service, because every step of the whole service process from negotiation to finally establishing customer relationship was carefully designed by Lexus Sales Department. For example, in negotiation, in order to make every customer delighted when selecting "our Lexus vehicles", Lexus Sales Department implemented a series of measures, for example, adopting order production system, exhibiting all models in Lexus stores, using negotiation assistant techniques and offering consistent insurance service.

As shown in Fig. 7.5, Lexus Sales Department, an explicit management innovation, affected the commercialization phase of product innovation through a series of implicit marketing strategy innovation.

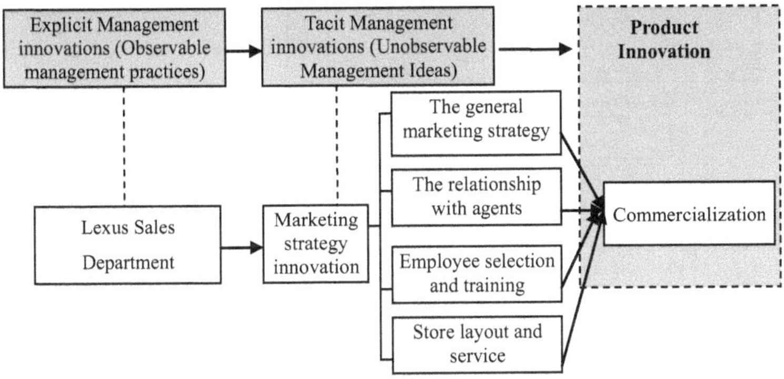

Fig. 7.5 Lexus Sales Department, marketing strategy innovation and product innovation

7.4.5 *Lexus Councils, Communication Mechanism Innovation and Product Innovation*

Through a corporate reorganization, Lexus was able to operate its own design, engineering, and manufacturing centers, solely responsible for Lexus vehicles, but it was a virtual structure which was not actually separated from Toyota Motor. Lexus division was integrated in the structure of Toyota Motor, though operated independently.

> At the very beginning of reorganization, I planned to set up a virtual Lexus company within the structure of Toyota Motor. That is, to establish director-level liaison meetings and corresponding councils to discuss and submit proposals, so as to realize horizontal communication. To do so, Lexus division could get enough support from various departments of Toyota Motor. (Vice president of Lexus division)

These meetings and councils were not offered with decision-making authorities. Their responsibility was to discuss certain issues and submit proposals to the business planning department for approval. For example, Lexus Liaison council, which horizontally covered departments of R&D, production, manufacturing, marketing and business planning of Toyota Motor, was set up to integrate the whole Lexus division together and make sure smooth internal communication; Lexus Marketing Council, which crossed business planning department, domestic and oversee marketing departments, was responsible for discussing reliability of product strategy and product models to make sure strategies workable; Lexus Product Council, which crossed departments of technology, production, manufacturing and business planning, attempted to control and improve the quality of Lexus products; Lexus Communication Council, which combined business planning department with marketing department, focused on improving the quality of Lexus service. In a word, this explicit innovation of setting up Lexus councils aimed to better integrate

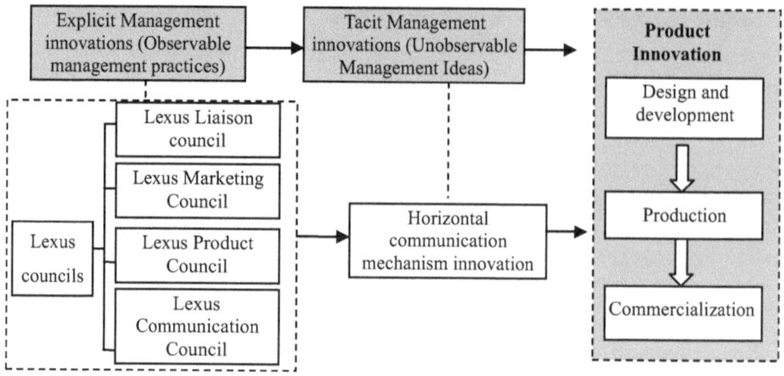

Fig. 7.6 Lexus councils, communication mechanism innovation and product innovation

Lexus division with Toyota Motor, to improve internal horizontal communication within the whole virtual Lexus division, and further to promote all product innovation activities, from design and development to commercialization, as shown in Fig. 7.6.

7.5 Discussion

7.5.1 The Framework of Management Innovation's Support to the Process of Product Innovation

Building on the exploration into the case of Lexus, a typical complex product with single brand, especially how management innovation supports each phase and even the whole process of Lexus product innovation during its reorganization period when rebuilding up the brand and launching it to the global market, this research sets up a framework to show the internal relationship between the two different types of innovation, as illustrated in Fig. 7.7.

Two groups of management innovation were identified: explicit management innovation and implicit management innovation. The former focused on observable management practices, namely structure changes in the case, and the latter referred to unobservable management ideas created by these newly-set-up structures. A virtual organization consisting of Business Reform Institution, Lexus (R&D) Center, Lexus Manufacturing Center, Lexus Sales Department, and Horizontal Councils was set up within the structure of Toyota Motor, but operated all product innovation activities independently when building a new single brand. The new structures or explicit management innovation themselves could not promote the process of product innovation, from design and development to production and finally to commercialization, but through a series of unobservable conception innovation or implicit management innovation. In the case of Lexus, implicit

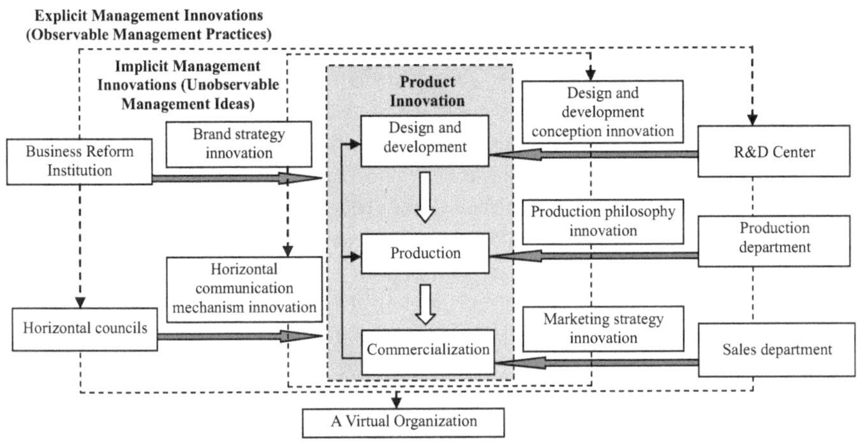

Fig. 7.7 The framework of management innovation's support to product innovation process

management innovation of brand strategy innovation, design and development conception innovation, production philosophy innovation, marketing strategy innovation and communication mechanism innovation were identified, which directly supported the process of Lexus product innovation. They were conducted by the five major new structures respectively.

When exploring deeply into how these explicit and implicit management innovation worked, this research finds that brand strategy innovation and horizontal communication mechanism innovation might affect all three phases of product innovation, while the other three each focused on a certain phase, for example, design and development conception innovation on new product design and development, production philosophy innovation on new product production, and marketing strategy innovation on new product commercialization. More specifically, brand strategy innovation conducted by Business Reform Institution might guide not only the whole process of product innovation by offering a separate brand position and enriching the concept of the new brand, but also the following explicit and implicit management innovation activities, so that the whole process of product innovation could stick to lofty standards of upscale and luxury cars. Design and development conception innovation conducted by Lexus Center referred to proposing a new design gospel, setting up special image strategy for each model of products and promoting continuous improvement and innovation. All these innovation activities aimed to better serve new product design and development. Production philosophy innovation made by Lexus Manufacturing Center stressed further lean manufacturing philosophy and a concept of Production Engineering Advanced Craftsmanship Effort, which focused on a pursuit for high quality, a target for elimination and an integration of advanced production technologies with superior craftsmanship. Without these management innovation activities, the new product innovation could not meet the lofty standards of luxury cars. Marketing strategy innovation carried out by Lexus Sales Department focused on offering new general marketing strategy, redefining the relationship with agents, changing employees selection and training mechanism, and creating the store layout and service process. All these activities aimed to better provide new products to customers and make them feel the most of every moment. Finally, setting up cross-department new brand Liaison Council, Marketing Council, Product Council and Communication Council helped to better horizontally communicate with parent company and get more support to the whole process of product innovation.

As shown in Fig. 7.7, all explicit management innovation or observable management practices make up a virtual organization when operating a new complex product innovation. Though this virtual organization serves as a foundation for the success of product innovation, it does not work without the support from a series of implicit management innovation or unobservable management ideas innovation. That is, explicit management innovation supports each phase of product innovation or even the whole process through implicit management innovation. Additionally, all explicit and implicit management innovation practices are interrelated and interdependent. In particular, Business Reform Institution guides all the other structure innovation, and brand strategy innovation directs all the other concept innovation.

7.5.2 *Implications for Theory*

This section focuses on the roles of management innovation in the process of complex product innovation through an exploratory case study on Lexus of Toyota Motor. It offers some implications to existing literature on management innovation and the alignment theory of management innovation and product innovation by addressing the two main questions mentioned in the introduction.

In consistence with Birkinshaw et al. (2008), this research advocates a separation of two levels of analysis in management innovation: an abstract level (also unobservable management ideas) and an operational level (also observable management practices). Most research on management innovation (e.g. Birkinshaw et al. 2008; Mol and Birkinshaw 2009; Alänge et al. 1998; Guillén 1994) focus on observable management practices at the operational level, "because this is the level at which observable changes take place in the way work is done and the management innovation process can be witnessed" (Birkinshaw et al. 2008, p. 828); some others prefer to management ideas, for example, an organizational ideology (Guillén 1994), and notions of management rhetoric (Barley and Kunda 1992; Abrahamson 1996; Suddaby and Greenwood 2005). Considering advantages of each level, this research combines both through a deep and longitudinal interpretive and exploratory case study on Lexus of Toyota Motor. In particular, this research uses mechanisms of archival data, semi-structured interviews and observation to collect abundant information on management innovation and then sort them into two groups of explicit management innovation (i.e., observable management practices) and implicit management innovation (i.e., unobservable management ideas). This research also finds that, the two groups of management innovation may coexist in the firm interdependently, and explicit management innovation support product innovation through implicit management innovation.

Prior studies have identified sequential phases through which management innovation occurs (e.g. Hamel 2006; Birkinshaw et al. 2008; Lin and Su 2014), examined conditions giving rise to the emergence of management innovation (e.g. Guillén 1994; McCabe 2002; Chi et al. 2007), and explored how to improve management innovation performance (e.g. Stata 1989; Garvin 1993; Teece 1997) and how management innovation is delivered (e.g. Abrahamson 1996; Teece 1980). However, few research have been done to explore how management innovation would contribute to organizational performance, though most studies on management innovation advocate a positive effect of management innovation on organizational performance and see it as one of the most important and sustainable sources of competitive advantage for firms (Hamel 2006; Mol and Birkinshaw 2009). Considering that management innovation is typically tacit in nature, and always full of uncertainty and ambiguity, this research argues that, it may contribute to organizational performance through more tangible product innovation, especially in firms producing complex products with a separate brand. This research on the case of Lexus confirms this view, and finds that both explicit and implicit management innovation are critical for the success of product innovation.

In consistence with the alignment theory of management innovation and product innovation which advocates an interrelationship between the two most important innovations in an organization (e.g. Vickery et al. 1999; Daft 1978; Xu and Xie 2004), this research suggests a support of management innovation to the process of product innovation and further answers the question of how. That is, this research stretches the alignment theory by exploring deeply into the effect of management innovation on product innovation. The research identifies two groups of management innovation (i.e., explicit management innovation and implicit management innovation), and finds that explicit innovation may produce indirect effects on product innovation, while implicit innovations may produce direct effects. Or, explicit innovations affect the process of product innovation through implicit innovations. More specifically, the three major phases of complex-separate-brand-product innovation, each needs the support from different management innovation. For example, in the phase of design and development, a new designing conception is needed to make new product distinctive; in the phase of production, a new manufacturing philosophy is needed to guarantee the quality of new products; and in the phase of commercialization, a new marketing strategy is desired to realize the value of new products and launch them to the market. In addition, some management innovation, such as brand strategy innovation and communication mechanism innovation, may support the whole process of product innovation.

7.5.3 Implications for Practice

Adopting a longitudinal interpretive and exploratory case study and blending management innovation, product innovation and alignment theory literature enable development of a general framework that explicates support and promotion mechanism of management innovation to the process of complex product innovation. More specifically, this work identifies major management innovation when operating a new complex product brand, divides them into two group of explicit (observable management practices) and implicit (unobservable management ideas) management innovation, and analyzes how these innovations support each phase of complex product innovation or even the whole process. This framework suggests a number of important insights for practice.

As product innovation treated as a weapon for the increase of firms' profits and a powerful factor behind differences in firms' performance, it has attracted abundant of scholars and practitioners to focus on the whole process of product innovation from design and development to production and commercialization. However, this research indicates that the success of product innovation, especially complex product with a separate brand, depends not only on the three phase themselves, but more on these management innovation which could support each phase. When joining the wave of product innovation to win the market, a large number of firms, especially middle and small ones, tend to simplify product innovation as a project of developing technologies with relative advantages and take actions blindly. They do not realize that it's a huge project concerning both management and technology capabilities of an organization. Especially when a firm attempts to build up a separate brand of a complex product (for example, automobiles), it should consider more about the system of innovation or alignment of management and technology. Therefore, this research shows how management innovation supports the process of complex product innovation to remind firms of the essential roles of management innovation with uncertain and lagging outcomes in launching a new product, and to offer key connecting points between two types of innovations. Additionally, when observable management practices has been proved critical for product innovation and even long-term success of firms, more attention should be given to innovation of unobservable management ideas or concept through which the explicit ones could work.

This research has answered the question that why Toyota Motor which had dedicated in manufacturing low-and-medium-scale vehicles for over 50 years before 1980s was able to successfully launch the best-selling luxury brand of Lexus within 30 years. It offers some valuable insights for automobile manufacturers or other complex products makers around the world especially those with similar culture and background who also desire to launch high-end products to the upscale market. Firstly, when building up a new separate brand of complex product, firms need to realize that the separation degree of new product from existing products depends on brand strategy innovation which may guide the whole process of product innovation and following management innovation, instead of technologies.

Secondly, all three sequential phases of product innovation need support of design and development conception innovation, production philosophy and strategy innovation and marketing strategy innovation. Accordingly, a cause for the failure of launching a new complex product with separate brand might be a lack of any implicit management innovation or explicit management innovation besides product innovation activities themselves. This confirms the theory of planning behavior that intentions determine actions (Ajzen 1985). Thirdly, it is impossible for firms to set up an actual new structure for development of a new complex product each time, so the process of product innovation is not completely separate from the origin structure, but still needs support from it. Therefore, horizontal communication mechanism needs to be set up to integrate new product innovation system into the whole structure of firms.

7.6 Conclusion

This section addresses what major management innovation may be conducted by a firm when building up a new brand of complex product like automobiles, and how these management innovation support each phase of product innovation or even the whole process from design and development, to production and to commercialization, aiming to offer an effective path for producing effects of management innovation on organizations. Based on existing literatures on management innovation, process of product innovation and the alignment between the two, five explicit management innovation and five implicit management innovation were identified in the case of Lexus, and a framework showing how these explicit management innovation support the product innovation process through implicit management innovation was set up. The findings offer many valuable insights for further research on how to explore the alignment mechanism between the two important innovations, and how management innovation with intangible, uncertain, lagging and even inseparable outputs contributes to performance of organizations, and hold important implications for management practices.

References

Abrahamson, E.: Management fashion. Acad. Manag. Rev. **21**(1), 254–285 (1996)

Ajzen, I.: From Intentions to Actions: A Theory of Planned Behavior. Springer, Berlin (1985)

Alänge, S., Jacobsson, S., Jaryehammar, A.: Some aspects of an analytical framework for studying the diffusion of organizational innovations. Technol. Anal. Strateg. Manag. **10**(1), 3–22 (1998)

Andriopoulos, C., Lewis, M.W.: Exploitation-exploration tensions and organizational ambidexterity: managing paradoxes of innovation. Organ. Sci. **20**(4), 696–717 (2009)

Barley, S.R., Kunda, G.: Design and devotion: surges of rational and normative ideologies of control in managerial discourse. Adm. Sci. Q. **37**(3), 363–399 (1992)

Bayus, B.L., Speed, G.: Design and devotion: surges of rational and normative ideologies of control in manage. **14**(6), 485–497 (1997)

Birkinshaw, J., Hamel, G., Mol, M.J.: Management innovation. Acad. Manage. Rev. **33**(4) (2008)

Calantone, R.J., Cooper, R.G.: A discriminant model for identifying scenarios of industrial new product failure. J. Acad. Mark. Sci. **7**(3), 163–183 (1979)

Calantone, R.J., Cooper, R.G.: New product scenarios: prospects for success. J. Mark. **45**(2), 48–60 (1981)

Calantone, R.J., Stanko, M.A.: Drivers of outsourced innovation: an exploratory study. J. Prod. Innov. Manag. **24**(3), 230–241 (2007)

Chi, W., Freeman, R.B., Kleiner, M.M.: Adoption and Termination of Employee Involvement Programs. NBER working paper, W12878, (2007)

Clark, K.B., Fujimoto, T.: Product Development Performance. Harvard Business School Press, Boston (1991)

Cooper, R.G.: Winning at New Products: Accelerating The Process from Idea to Launch, 2nd edn. Addison-Wesley, Reading, MA (1993)

Cooper, R.G., Edgett, S.C.: Maximizing productivity in product innovation. Res. Technol. Manage. **51**(2), 47–58 (2008)

Cooper, R.G.: Perspective: the stage-gate® idea-to-launch process—update, what's new, and nexgen systems. J. Prod. Innovation Manage. **25**(3), 213–232 (2008)

Cooper, R.G., Edgett, S.J., Kleinschmidt, E.J.: New product portfolio management: practices and performance. J. Prod. Innov. Manag. **16**(4), 333–351 (1999)

Cooper, R.G., Edgett, S.J., Kleinschmidt, E.J.: Benchmarking best NPD practices—the NPD process and key idea-to-launch activities. Res. Technol. Manage. **47**, 43–55 (2005)

Corsino, M., Gabriele, R.: Product innovation and firm growth: evidence from the integrated circuit industry. Ind. Corp. Change **20**(1), 29–56 (2011)

Crawford, C.M.: Defining the charter for product innovation. Sloan Manag. Rev. **21**(3), 3–12 (1980)

Czarniawska, B., Sevón, G.: Translating Organizational Change. W. de Gruyter (1996)

Czarniawska, B., Sevón, G.: Global Ideas: How Ideas, Objects and Practices Travel in A Global Economy. Copenhagen Business School Press, Denmark (2005)

Daft, R.L.: A dual-core model of organizational innovation. Acad. Manag. J. **21**(2), 193–210 (1978)

Deetz, S., Tracy, S.J., Simpson, J.L.: Leading Organizations Through Transition: Communication and Cultural Change. Sage Publications, Inc. (2000)

Eisenhardt, K.M.: Building theories from case study research. Acad. Manag. Rev. **14**(4), 532–550 (1989)

Eisenhardt, K.M., Graebner, M.E.: Theory building from cases: opportunities and challenges. Acad. Manag. J. **50**(1), 25–32 (2007)

Garvin, D.A.: Building a learning organization. Harvard Bus. Rev. **71**(4), 78–91 (1993)

Glaser, B.G., Strauss, A.L.: The Discovery of Grounded Theory: Strategies for Qualitative Research. Aldine, Chicago (1967)

Golden-Biddle, K., Locke, K.: Composing Qualitative Research: Crafting Theoretical Points from Qualitative Research. Sage, CA (2007)

Griffin, A.: Metrics for measuring product development cycle time. J. Prod. Innov. Manag. **10**, 112–125 (1993)

Guillén, M.F.: Models of Management: Work, Authority, and Organization in A Comparative Perspective. University of Chicago Press, Chicago (1994)

Gummesson, E.: Qualitative Research Methods in Management Research. Sage, Thousand Oaks, CA (2000)

Hamel, G.: The why, what, and how of management innovation. Harvard Bus. Rev. **84**(2), 72–84 (2006)

Henard, D.H., Szymanski, D.M.: Why some new products are more successful than others. J. Mark. Res. **38**(3), 362–375 (2001)
Henderson, R.M., Clark, K.B.: Architectural innovation: the reconfiguration of existing product technologies and the failure of established firms. Adm. Sci. Q. **35**(1), 9–30 (1990)
Hox, J.J., Boeije, H.R.: Data collection, primary versus secondary. Encycl. Soc. Meas. **1**, 593–599 (2005)
Hultink, E.J., Griffin, A., Hart, S., Robben, H.S.J.: Industrial new product launch strategies and product development performance. J. Prod. Innov. Manag. **14**(4), 243–257 (1997)
Krippendorff, K.: Content Analysis: An Introduction to Its Methodology, 2nd edn. Sage, Thousand Oaks. CA (2004)
Langley, A.: Strategies for theorizing from process data. Acad. Manag. Rev. **24**(4), 691–710 (1999)
Lin, H.F., Su, J.Q.: A case study on adoptive management innovation in China. J. Organ. Change Manage. **27**(1), 83–114 (2014)
Lincoln, Y.S., Guba, E.G.: Naturalistic Inquiry Sage. Sage Publications, Naturalistic inquiry (1985)
Mccabe, D.: 'Waiting for dead men's shoes': towards a cultural understanding of management innovation. Hum. Relat. **55**(5), 505–536 (2002)
Mcnally, R.C., Akdeniz, M.B., Calantone, R.J.: New product development processes and new product profitability: exploring the mediating role of speed to market and product quality. J. Prod. Innov. Manag. **28**(1), 63–77 (2011)
Miles, R.E., Snow, C.C.: Organizational strategy, structure, and process. McGraw-Hill, New York (1978)
Mol, M.J., Birkinshaw, J.: The sources of management innovation: when firms introduce new management practices. J. Bus. Res. **62**(12), 1269–1280 (2009)
Mol, M., Birkinshaw, J.: How management innovation happens. Mit Sloan Manage. Rev. **47**(4), 81–88 (2006)
Nayak, P.R.: Managing Rapid Technological Development. Arthur D. Little, Boston, MA (1991)
Nelson, R.R., Winter, S.G.: An evolutionary theory of economic change. Belknap Press/Harvard University Press, Cambridge (1982)
OECD, P.E.: Technology and Productivity: The Challenge for Economic Policy (1991)
Pil, F.K., Macduffie, J.P.: The adoption of high-involvement work practices. Ind. Relat. J. Econ. Soc. **35**(3), 423–455 (1996)
Sethi, R.: New product quality and product development teams. J. Mark. **64**(2), 1–14 (2000)
Smith, P.G., Reinertsen, D.G.: Developing Products in Half The Time: New Rules. Van Nostrand Reinhold, New Tools. John Wiley. New York (1998)
Stata, R.: Organizational learning-the key to management innovation. Mit Sloan Manage. Rev. **63** (1), 63–73 (1989)
Suddaby, R., Greenwood, R.: Rhetorical strategies of legitimacy. Adm. Sci. Q. **50**(1), 35–67 (2005)
Swink, M.: Technological innovativeness as a moderator of new product design integration and top management support. J. Prod. Innov. Manag. **17**(3), 208–220 (2000)
Teece, D.J.: The diffusion of an administrative innovation. Manage. Sci. **26**(5), 464–470 (1980)
Teece, D.J., Pisano, G., Shuen, A.: Dynamic capabilities and strategic management. Strateg. Manag. J. **18**(7), 509–533 (1997)
Tidd, J., Bessant, J., Pavitt, K.: Managing Innovation Integrating Technological Market and Organizational Change, 2nd edn. Wiley, Chichester (2001)
Tzokas, N., Hultink, E.J., Hart, S.: Navigating the new product development process. Ind. Mark. Manage. **33**(7), 619–626 (2004)
Utterback, J.M.: Mastering The Dynamics of Innovation. Harvard Business School Press, Boston (1994)

Vaccaro, I.G., Jansen, J.J.P., Bosch, F.A.J.V.D., Volberda, H.W.: Management innovation and leadership: the moderating role of organizational size. J. Manage. Stud. **49**(1), 28–51 (2012)
Vickery, S., Dröge, C., Germain, R.: The Relationship between product customization and organizational structure. J. Oper Manage. **17**(4), 377–391 (1999)
Xu, Q.R., Xie, Z.S.: Study of firm's innovation synergy and the evolvement model. Stud. Sci. Sci. **22**, 327–332 (2004)
Yin, R.K.: Case Study Research: Design and Methods, 2nd edn. Sage, Newbury Park, CA (1994)

Chapter 8
Conclusion

8.1 Findings

Considering the importance and popularity of adoptive management innovation in China and even around the world, and focusing on its nature, this research attempts to explore the adoptive management innovation and set up a theory on this field. It sets up a process framework through which adoptive management innovation occurs, explores how individual characteristics of core individual managers affects their adoption decision and how firms implement management innovation deeply into the most micro level of organizational routines, examines the effects of dynamic capability of a firm on each phase of adoptive innovation, and addresses how intangible management innovation supports the process of tangible product innovation to produce effects. Six major findings could be obtained from this research.

(1) A two-subprocess framework consisting of adoption decision and implementation for adoptive management innovation was built up.

The results of an exploratory case study on Organizational Efficiency Management innovation of Jiangxi Mobile in China indicate that the process of adoptive management innovation consists of adoption decision and implementation. Adoption decision refers to activities of problems identification, innovation perception, attitude formation, problem diagnoses, innovation revision, proposal evaluation and yes-no selection; and implementation reflects events and actions that pertain to modifying the innovation, preparing the organization for its use, trial use, acceptance of the innovation and continued use of the innovation. The results indicate that the adoption of existing management practices or methods from somewhere else is a more complex and logical process rather than a simple one of knowledge transferring. It needs to integrate existing practices into new organizational context and establish their innovative value during implementation. One core element of the process framework is the emphasis on activities of problem

H. Lin, *Adoptive Management Innovation*,
https://doi.org/10.1007/978-981-10-7712-8_8

diagnoses and realization of the fitness between management practices adopted and the new organizational context, and another one is the sequence of activities in the whole process.

(2) A three-dimension decision framework was set up and confirmed through SEM.

The results indicate that innovation intention, knowledge acquirement and risk perception of individual managers directly affect management innovation decision level; innovation intention, knowledge acquirement and risk perception are influenced by entrepreneurial orientation, social network and cognitive biases of managers respectively; besides the three relevantly independent affecting paths, interrelations also exist cross paths; the three main factors produce both direct and indirect effect simultaneously.

(3) A three-phase evolution process of organizational routines in management innovation implementation was set up through a case study.

The research confirms the management innovation is a complex project concerning organizational routines which represent the most micro feature of an organization. The success of DD innovation in HOAU relies not only on setting up a good innovation proposal covering a new structure, a new process and a set of new criterions, but importantly on the effective implementation which reflects a process involving all organization levels and especially a fundamental change of the implicit routines. The evolution of organizational routines occurs through a process consisting of existing routines domination, new routines creation and new routines solidification, each exhibiting different innovation activities and characteristics of participants' cognition and behaviors. Moreover, the recreations of new routines are the keys for routine evolution, thus for the success of management innovation. If an innovation stops at breaking existing routines rather than creating new routines, it would fail to yield the intended results, even make the organization trapped into a new development dilemma.

(4) A measuring scale for adoptive management innovation with 14 items (2 items deleted) was set up by adopting a process-oriented method rather than a result-oriented one.

The reliability and validity assessment showed that all of the indicators had individual indicator reliability values that were larger than 0.7, except for Out_1 (i.e., timing of existing solution acquisition) for outside search and Pro_2 (i.e., timing of reestablishment,) for proposal establishment, which confirms a well-developed scale for measuring adoptive management innovation. The adopting of the process-oriented method and the setting up of the 14-item measuring scale may further research on adoptive management innovation and offer valuable insights for measuring generating innovation and other complicated organization behaviors.

(5) The positive effects of dynamic capabilities on the whole process of adoptive management innovation were generally confirmed.

PLS-SEM was adopted to examine both the structural model reflecting the relationships between four components of dynamic capabilities and four phases of adoptive management innovation (which was reset up by holding an organizational perspective in Sect. 8.4) and the measurement model reflecting indicators of each construct. The result reveals the strongest total effect of sensing capability on outside search, followed by initiation and implementation, but not proposal establishment; the strongest effect of absorptive capacity on implementation, followed by proposal establishment, outside search and initiation; the most important role of integrative capability in supporting implementation, followed by initiation and proposal establishment, but not outside research; positive and significant total effects of relational capability on all four phases of adoptive management innovation, with the strongest effect on implementation, followed by outside search, initiation and proposal establishment. In a word, each component affects each phase with unique path, and relational capability based on relationships with firms and individuals serves as a fundamental role, which reflects the culture in China where affection rooting in relationships determines acquisition of resources and even survival and development of organizations.

(6) A general framework that explicates support and promotion mechanism of management innovation to the process of complex product innovation was set up.

A longitudinal interpretive and exploratory case study on Lexus of Toyota Motor indicates that intangible management innovation may affect organizational performance through intangible product innovation. More specifically, both explicit management innovation (i.e., observable management practices) and implicit management innovation (i.e., unobservable management concepts) may interdependently coexist in the firm when creating a complex product with a separate brand, and the former may support the process of product innovation through the latter.

8.2 Implications for Theory

This research, focusing on the process, adoption decision, implementation, driver and effects of adoptive management innovation, offers various implications to existing literature on management innovation by addressing the four questions mentioned in the introduction.

In consistence with the studies of Mol and Birkinshaw (2009) and Vaccaro et al. (2012), but contrary to that of Birkinshaw and Mol (2006), Birkinshaw et al. (2007, 2008) who give more attention to new-to-the-state-of-the-art innovation, namely generating management innovation, this research suggests a focus on new-to-the-organization innovation, namely adoptive management innovation, or an application of management innovation to the organizational level of analysis. It expands the research of management innovation to a larger scale by covering deeper micro-level organizational behaviors, and further stresses the value of focusing on new-to-the-organization innovation by following Mol and Birkinshaw (2009) and Vaccaro et al. (2012). Moreover, in order to catch up with international developed firms in management, a large number of firms especially less-developed Chinese firms have engaged in learning, adopting and implementing management innovation practices already proved to be effective or successfully implemented in other organizations with relevantly low risk, in comparison with an unprecedented change with higher risk and uncertainty. It makes this research much closer to the common-existing practices of domestic firms and even firms all over the world. In another word, a deep investigation into new-to-the-organization innovation could offer more effective guidance, though less new knowledge.

This research reflects the argument made by Birkinshaw and Mol (2006), Birkinshaw et al. (2007, 2008) that the generative mechanisms of management innovation through which it occurs are theoretically interesting in their own right and also relatively poorly understood. Birkinshaw et al. (2008) develop a framework based on new-to-the-state-of-the-art management innovation highlighting four interlinked phases of motivation, invention, implementation and theorization and labeling, while this research proposes a general process with two key interlinked subprocesses of adoption decision and implementation for adoptive management innovation, each containing various activities. Besides some common findings, for example, both stress new problem identification in the early period of innovation and reveal that implementation is an indispensable phase, this research on adoptive management innovation differs on four aspects at least, which indicates to some degree the differentiation between the two types of management innovation. For example, adoptive innovation is introduced by internal agents from somewhere else instead of being created by a firm; instead of the activity of trial and error in which progress is achieved by monitoring and making adjustments against the original concept, this research emphasizes activities of problem diagnoses and realization of the fitness between introduced management practice and the new organizational context; the framework on generative mechanism of completely new innovation stresses the important roles of both internal and external change agents in the process, while this research focuses on internal agents only, especially the role of core manager in adoption decision; theorization and labeling, the fourth phase in the framework of generating innovation, was not mentioned in this research.

Research literatures particular those on the rationality theory stress the important role of individual managers in adopting and implementing new management practices. They argue that it is managers that address critical problems by creating or adopting new practices and offering support for implementation (Howell and Higgins 1990). This study contributes to these researches by proposing that entrepreneurial orientation, social network and cognition of managers are important in understanding the introduction of new practices. By showing specific affecting paths of three major characteristics of managers, three main activities abstracted from the practices and literatures of management innovation, this research aims to unveil the black box in complex decision process.

This study is contrary to most researchers (e.g. Mol and Birkinshaw 2009; Vaccaro et al. 2012) who adopt a result-oriented method for measuring management innovation. Since the results of management innovation are always intangible, uncertain, lagging and even inseparable from that of technological innovation in organizations (Birkinshaw et al. 2008), which indicates a difficulty in quantifying the results, a process-oriented method for exploring deeply into the nature of management innovation was advocated. Taking insights from the four-phase process of generating management innovation set up by Birkinshaw et al. (2008), and especially the two-general-phase process of adoptive innovation set up in Sect. 8.3, this research further proposes a four-phase process consisting of initiation, outside search, proposal establishment and implementation for adoptive management innovation in Sect. 8.4. More importantly, a scale with 14 items (2 items deleted) was developed based on an extraction of characteristics and major activities in each phase of innovation, which may further research on adoptive management innovation and offer valuable insights for measuring generating innovation and other complicated organization behaviors.

Also, this research extends research on management innovation to most micro level by focusing on innovation implementation. It particularly contributes to research on the relationship between organizational routines and innovation by taking an organizational routines' perspective to explore management innovation and drawing the conclusion that implementation of management innovation reflects the evolution of organizational routines. Prior research on innovation has regarded innovation as a systematic project potentially requiring fundamental changes in the routines of the organization (e.g. Argyris and Schän 1978; Birkinshaw et al. 2008). At the same time, recent research on organizational routines has regarded organizational routines as the fundamental to understanding organizational changes or innovations (Becker et al. 2005), or as an important source of flexibility, change and innovation (Nelson and Winter 1982; Pentland and Rueter1994; Feldman 2000; Feldman and Pentland 2003, 2008; Pentland and Feldman 2005; Pentland et al. 2011, 2012). Consistently, this research confirms the relationship between organizational routines and innovation and argues that the success of management innovation relies on evolution of organizational routines. Following Becker et al. (2005), this research advocates the necessary of exploring into the micro-level change in the innovation process. The result indicates that the implicit perspective in management innovation reflects the underlying change of participants' cognition and behaviors, namely the change of shared schemata and action dispositions, or that of organizational routines, a central, micro and fundamental element of organizations (Feldman and Pentland 2003; March and Simon 1958; Cyert and March 1963; Cohen and Klepper 1996; Ortmann 2010). Accordingly, antecedents affect innovation through cognition and behaviors of employees and then evolution of routines. This explains why organizational routines are sources of innovations. Moreover, this research furthers research on organizational routine theory by advocating the two-component view and investigating deeply into the process through which organizational routines evolve.

Prior studies have examined effects of different antecedents on management innovation, for example, social economic conditions (Guillén 1994; Kossek 1987), external pressure (DiMaggio and Powell 1983), key individuals within organizations (Howell and Higgins 1990; Vaccaro et al. 2012), cultural condition (McCabe 2002) and human resources (Osterman 1994; Chi et al. 2007). However, this research argues that management innovation rely more on the whole organizational system with dynamic capabilities of integrating, reconfiguring, gaining and releasing resources. In consistence with most studies on the relationship between dynamic capability and innovation (e.g. Teece et al. 1997, 2007; Helfat et al. 2007; Ridder 2011; Aragon-Correa and Sharma 2003; Hart and Dowell 2010; Clausen 2013; Ambrosini et al. 2009; Kohlbacher 2013; Cheng and Chen 2013), this research confirms positive effects of dynamic capability on the process of adoptive management innovation by advocating a multidimensional construct view of dynamic capability (e.g. Teece 2007; Wang and Ahmed 2007; Barreto 2010), instead of a single-dimensional construct (e.g. Liao et al. 2009; Cheng and Chen 2013). Four distinctive but related components of dynamic capability referring to sensing capability, absorptive capacity, relational capability, and integrative capability were identified through a Fuzzy cluster analysis. This attempt may further research on dynamic capability and its component identification. PLS-SEM was adopted to examine both the structural model reflecting the relationships between four components of dynamic capability and four phases of adoptive management innovation and the measurement model reflecting indicators of each construct. To the author' expectation, the proposed effects of dynamic capability on adoptive management innovation were generally confirmed, though not completely corresponding.

Though most studies on management innovation advocate a positive effect of management innovation on organizational performance and see it as one of the most important and sustainable sources of competitive advantage for firms (Hamel 2006; Mol and Birkinshaw 2009), few research have been done to explore how management innovation would contribute to organizational performance. Considering that management innovation is typically tacit in nature, and full of uncertainty and ambiguity, this research argues that, it may contribute to organizations by directly servicing more tangible product innovation, especially complex products with a separate brand. In consistence with the alignment theory of management innovation and product innovation which advocates an interrelationship between the two most important types of innovation in an organization (e.g. Vickery et al. 1999; Daft 1978; Xu and Xie 2004), this research suggests a support of management innovation to the process of product innovation and further address the problem of how. In another word, this research stretches the alignment theory by explore deeply into effect of management innovation on product innovation or the dependence of the latter on the former. The result indicates that, among the two groups of management innovation identified, explicit innovations may produce indirect effect, while implicit innovations may produce direct effect. Or explicit innovations affect the process of product innovation through implicit innovations. As for the three major phases of complex-separate-brand product innovation, each needs support from different kinds of management innovation.

8.3 Applications for Practice

The exploration on the process, adoption decision, driver and effects of adoptive management innovation in this research may offer the following applications for managerial practice:

The built-up process framework that explicates realization mechanism of adoptive management innovation indicates that though adoptive management innovation indicates introduction and implementation of management practices implemented successfully somewhere else, it is not a simple job of repetition, but still a huge project with a complicated process full of uncertainty. When joining the wave of management innovation, a large number of firms, especially middle and small Chinese firms, underestimate the risk of adoptive management innovation and take actions blindly. They realize the potential of adoptive innovation in improving management and even performance of the whole firm, but neglect the accompanying risk and uncertainty. Therefore, this research shows how adoptive management innovation comes about to remind firms of risks and difficulties.

By resetting up a framework of adoption decision and showing specific affecting paths of entrepreneurial orientation, social network and cognition of managers through innovation intention, knowledge acquirement and risk perception respectively to adoption level of management innovation, this research confirms adoption process a complex process. More specifically, managers with stronger innovation intention, higher levels of knowledge acquirement and higher levels of risk perception, tend to be more active in introducing new practices. And further, entrepreneurial orientation, social network and cognition of managers significantly affect their entrepreneurial orientation, social network and cognition, and subsequently adoption of new practice. That is, it reminds managers of a firm the importance of their entrepreneurial orientation, social network and cognition biases in adopting new management practices.

As this research shows management innovation is a systematic project referring to evolution of organizational routines, it is in the interest of firms to call attention of all participants, not only core managers. Besides phases of initiation, decision making and proposal establishment in innovation which rely on top managers to a large extent, innovation implementation is an indispensible phase involving participants even from the first line. Why many firms set up perfect proposals or adopt complete management practices from somewhere else but are still not able to produce intended results may be a lack of attention paid to implementation. Therefore, this research stresses the value of implementation to remind managers that implementation is not a naturally-coming phase. More importantly, besides the explicit changes of the organizational structure, the process or institutions, firms should carefully observe accompanying changes in cognition and behaviors of participants engaged in innovation which reflects the implicit evolution of organizational routines. Only when existing organizational routines are overturned and new routines are created and solidified to further guide participants could we judge that the innovation has been successfully implemented. This reminds managers of the complication of innovation implementation. Though the evolution of organizational routines in innovation is an endogenous process proceeding through interactions and role taking, it is necessary for firms to take actions to appropriately stimulate interactions and role taking among participants, so as to shorten the period of evolution and better realize the goal of innovation.

With difficulty in quantifying outputs of organizational practices, a process perspective may be adopted in assessing their performance. As this research indicates, the number of new practices initiated, attention paid to renewing management practices, management problems identification and timing of initiation are good indicators for performance of initiation; acquaintance with emerging practices in market sources, internal sources and professional sources are good indicators for performance of outside search; the preference on creating a new proposal, match of new established proposal with the original one and quality of proposals are good indicators for performance of proposal establishment; and the ratio of proposals taken into practice, the support of the whole organization, progress of implementation and outputs of innovation are good indicators for innovation implementation.

This research offers a dynamic capability perspective by exploring and confirming effects of its four components on each phase of adoptive management innovation, which confirms dynamic capability an important driver of management innovation. The results indicate that management innovation is not isolated, but a systematic project depending on all the four components of dynamic capabilities. More specifically, strong sensing capability may be able to directly or indirectly stimulate activities of outside search (with the strongest total effect), initiation and implementation. However, sensing capability is not able to improve performance of proposal establishment. Similarly, the research confirms positive and significant total effects of absorptive capacity on all four phases of innovation, especially on implementation, which reminds managers of the necessity of offering employees with knowledge acquiring, processing, combining and creating guidelines and tools. Integrative capability may be able to directly or indirectly support activities of implementation, proposal establishment and initiation. However, integrative capability does not significantly impact outside search due to its weak effect in the linkage. Finally, relational capability is confirmed as an important indicator of other three capacities. Though it may not be able to produce direct effect on proposal establishment, it does affect proposal establishment through other capabilities or previous phases of innovation. As a result, much attention should be paid to set up close and trusted relationships with other firms, government and institutions and individuals as well as internal employees to support process of complicate organizational changes like management innovation to finally improve organizational performance.

This research proposes a path from management innovation to organizational performance, namely through product innovation. It confirms that intangible management innovation may affect organizational performance through tangible product innovation. More specifically, this work identifies major management innovation when operating a new complex product brand, divides them into two groups of explicit (observable management practices) and implicit (unobservable management ideas) management innovation, and analyzes how these innovations support each phase of complex product innovation or even the whole process. In another word, this research indicates that the success of product innovation, especially complex product with a separate brand, depends not only on the three phase themselves, but more on these management innovation which could support each phase. When joining the wave of product innovation to win the market, a large number of firms tend to simplify product innovation as a project of developing technologies with relative advantages and take actions blindly. They do not realize that it's a huge project concerning both management and technology capabilities of an organization. Especially when a firm attempts to build up a separate brand of a complex product (for example, automobiles), it should consider more about the system of innovation or alignment of management and technology. Therefore, this research shows how management innovation supports the process of complex product innovation to remind firms of the essential roles of management innovation with uncertain and lagging outcomes in launching a new product, and to offer key connecting points between two types of innovations. Additionally, when observable management practices has been proved critical for product innovation and even long-term success of firms, more attention should be given to innovation of unobservable management ideas or concept through which the explicit ones could work.

8.4 Limitations and Future Directions

This effort towards identifying process, decision mechanism, implementation, driver and effect of adoptive management innovation at the firm level is constrained by three limitations, which also represents fertile ground for future research in this area.

First, the findings for the two-subprocess framework of the adoptive management innovation, the decision making process, the evolution of organizational routine in innovation implementation, the effects of management innovation on performance of an organization, need to be further confirmed for more organizations. That is, more cases of adoptive management innovation should be adopted and compared to improve these frameworks and further explore into details.

Second, the research does not investigate other variables that may affect innovation process, innovation decision, and innovation implementation, such as contextual elements like competitive intensity, internal elements like structure and culture, as well as top management teams.

Third, though this research had focused on four components of dynamic capabilities, there may be some others not mentioned, for example, seizing capability (Teece 2007), adaptive capability (Wang and Ahmed 2007), or learning capability.

Additionally, the data were collected by questionnaires, so further longitudinal research using observation and interviews could contribute to this area by providing a richer understanding.

In addition, this research does not address other relevant problems, for example, why some managers pursue innovative opportunities by introducing new management practices while others do not; how dynamic capabilities affect the process of generating management innovation; how resources of firms which service as the base for dynamic capabilities affect the formation of dynamic capabilities and even management innovation; how product innovation affects management innovation in a firm; or how other elements of firms (for example, competitive intensity, structure, culture, or top management teams) affect the process of management innovation, especially in the context of China. Therefore, future research may be concerned with these issues.

References

Ambrosini, V., Bowman, C., Collier, N.: Dynamic capabilities: an exploration of how firms renew their resource base. Brit. J. Manag. **20**(Supplement s1), S9–S24 (2009)

Aragón-Correa, J.A., Sharma, S.: A contingent resource-based view of proactive corporate environmental strategy. Acad. Manag. Rev. **28**(1), 71–88 (2003)

Argyris, C., Schön, D.A.: Organizational learning: a theory of action perspective. Birkhäuser Basel (1978)

Barreto, I.: Dynamic capabilities: a review of past research and an agenda for the future. J. Manag. **36**(1), 256–280 (2010)

Becker, M.C., Lazaric, N., Nelson, R.R., Winter, S.G.: Applying organizational routines in understanding organizational change. Post-Print **14**(5), 775–791 (2005)

Birkinshaw, J., Crainer, S., Mol, M.: Special report on management innovation. Bus. Strategy Rev. **18**(1), 45–73 (2007)

Birkinshaw, J., Hamel, G., Mol, M.J.: Management innovation. Acad. Manag. Rev. **33**(4) (2008)

Mol, M., Birkinshaw, J.: How management innovation happens. Mit Sloan Manag. Rev. **47**(4), 81–88 (2006)

Cheng, C.C.J., Chen, J.S.: Breakthrough innovation: the roles of dynamic innovation capabilities and open innovation activities. J. Bus. Ind. Mark. **28**(5), 444–454 (2013)

Chi, W., Freeman, R.B., Kleiner, M.M.: Adoption and termination of employee involvement programs. NBER Working Paper, W12878 (2007)

Clausen, T.H.: The Role of Operational and Dynamic Capabilities in Ambidextrous Innovation: A Longitudinal Analysis. In. (2013)

Cohen, W.M., Klepper, S.: Firm size and the nature of innovation within industries: the case of process and product R&D. Rev. Econ. Stat. **78**(2), 232–243 (1996)

Cyert, R.M., March, J.G.: A Behavioral Theory of The Firm. Prentice-Hall, Englewood Cliffs, NJ (1963)

Daft, R.L.: A dual-core model of organizational innovation. Acad. Manag. J. **21**(2), 193–210 (1978)

DiMaggio, P.J., Powell, W.W.: The iron cage revisited: institutional isomorphism and collective rationality in organizational fields. Am. Sociol. Rev. **48**(2), 147–160 (1983)

Feldman, M.S.: Organizational routines as a source of continuous change. Organ. Sci. **11**(6), 611–629 (2000)

Feldman, M.S., Pentland, B.T.: Reconceptualizing organizational routines as a source of flexibility and change. Adm. Sci. Q. **48**(1), 94–118 (2003)

Feldman, M.S., Pentland, B.T.: Designing routines: on the folly of designing artifacts, while hoping for patterns of action. Inf. Organ. **18**(4), 235–250 (2008)

Guillén, M.F.: Models of Management: Work, Authority, and Organization in A Comparative Perspective. University of Chicago Press, Chicago (1994)

Hamel, G.: The why, what, and how of management innovation. Harvard Bus. Rev. **84**(2), 72–84 (2006)

Hart, S.L., Dowell, G.: A natural-resource-based view of the firm: fifteen years after. J. Manag. **37** (5), 1464–1479 (2010)

Helfat, C.E., Finkelstein, S., Mitchell, W., Peteraf, M., Singh, H., Teece, D., Winter, S.G.: Dynamic capabilities: understanding strategic change in organizations. Blackwell, Oxford, U. K. (2007)

Howell, J.M., Higgins, C.A.: Champions of technological innovation. Adm. Sci. Q. **35**(2), 317–341 (1990)

Kohlbacher, M.: The impact of dynamic capabilities through continuous improvement on innovation: the role of business process orientation. Knowl. Proc. Manag. **20**(2), 71–76 (2013)

Kossek, E.E.: Human resources management innovation. Hum. Resour. Manag. **26**(1), 71–92 (1987)

Liao, J., Kickul, J.R., Ma, H.: Organizational dynamic capability and innovation: an empirical examination of internet firms. J. Small Bus. Manage. **47**(3), 263–286 (2009)

March, J.G., Simon, H.A.: Organizations. Wiley, New York (1958)

Mccabe, D.: 'Waiting for dead men's shoes': towards a cultural understanding of management innovation. Hum. Relat. **55**(5), 505–536 (2002)

Birkinshaw, J., Mol, M.J.: The sources of management innovation: when firms introduce new management practices. J. Bus. Res. **62**(12), 1269–1280 (2009)

Nelson, R.R., Winter, S.G.: An Evolutionary Theory of Economic Change. Belknap Press/Harvard University Press, Cambridge (1982)

Ortmann, G.: On drifting rules and standards. Scand. J. Manag. **26**(2), 204–214 (2010)

Osterman, P.: How common is workplace transformation and who adopts it? Ind. Labor Relat. Rev. **47**(2), 173–188 (1994)

Pentland, B.T., Feldman, M.S.: Organizational routines as a unit of analysis. Ind. Corp. Change **14** (5), 793–815 (2005)

Pentland, B.T., Feldman, M.S., Becker, M.C., Liu, P.: Dynamics of organizational routines: a generative model. Soc. Sci. Electron. Publishing **49**(8), 1484–1508 (2012)

Pentland, B.T., Rem, T., Hillison, D.: The (N)ever-changing world: stability and change in organizational routines. Organ. Sci. **22**(6), 1369–1383 (2011)

Pentland, B.T., Rueter, H.H.: Organizational routines as grammars of action. Adm. Sci. Q. **39**(3), 484–510 (1994)

Ridder, A.K.: Sensing and seizing open innovation: a capability-based approach. In: Paper presented at the DIME-DRUID ACADEMY Winter Conference 2011. Comwell Rebild Bakker (2011)

Teece, D.J.: Explicating dynamic capabilities: the nature and microfoundations of (sustainable) enterprise performance. Strateg. Manag. J. **28**(13), 1319–1350 (2007)

Teece, D.J., Pisano, G., Shuen, A.: Dynamic capabilities and strategic management. Strateg. Manag. J. **18**(7), 509–533 (1997)

Vaccaro, I.G., Jansen, J.J.P., Bosch, F.A.J.V.D., Volberda, H.W.: Management innovation and leadership: the moderating role of organizational size. J. Manage. Stud. **49**(1), 28–51 (2012)

Vickery, S., Dröge, C., Germain, R.: The relationship between product customization and organizational structure. J. Oper. Manag. **17**(4), 377–391 (1999)

Wang, C.L., Ahmed, P.K.: Dynamic capabilities: a review and research agenda. Int. J. Manag. Rev. **9**(1), 31–51 (2007)

Xu, Q.R., Xie, Z.S.: Study of firm's innovation synergy and the evolvement model. Stud. Sci. Sci. **22**, 327–332 (2004)

Appendix A
Questionnaire for Adoption Decision of Management Innovation

1. **Items of *adoption level of management innovation***

Did your company make major changes in the following areas of business practices during the period of 2007–2009?

(1) Implementation of new or significantly changed corporate strategies; (0 = no; 1 = yes)
(2) Using more advanced management techniques; (0 = no; 1 = yes)
(3) Changing the organizational structure; (0 = no; 1 = yes)
(4) Implementation of business process reengineering; (0 = no; 1 = yes)
(5) Reforming the organizational culture; (0 = no; 1 = yes)
(6) Changing significantly your firm's marketing concepts/strategies; (0 = no; 1 = yes)
(7) Introducing new methods of risk-controlling and financial management; (0 = no; 1 = yes)
(8) Implementation of new human resource managements or channel management systems; (0 = no; 1 = yes)
(9) Other new practices(0 = no; 1 = yes).

2. **Items of *entrepreneurial orientation***

(1) More attention has been paid to technological innovation;
(2) A large number of technological innovation projects have been promoted during the last three years;
(3) You prefer to radical changes;
(4) You initiate actions to which competitors then respond;
(5) You are always the first to introduce new products or management pattern;
(6) You adopt a very competitive, undo-the-competitors posture;
(7) There is a strong proclivity for high-risk projects;

H. Lin, *Adoptive Management Innovation*,
https://doi.org/10.1007/978-981-10-7712-8

(8) Owning to the nature of the environment, bold, wide-ranging acts are necessary to achieve the firm's objectives;
(9) You adopt a bold, aggressive posture in order to maximize the probability of exploiting potential opportunities.

All items were measured on a 5-item scale, on which "1 = strongly disagree" and "5 = strongly agree".

3. **Items of *innovation intention***

(1) You are considering introducing and implementing at least one new management practice now;
(2) The existing operation or management system of your organization is temporal and changeable;
(3) At least one practice will be introduced in the following year;
(4) The idea of changing the existing operation or management system always exists;
(5) The existing operation or management system needs to be improved dramatically.

All items were measured on a 5-item scale, on which "1 = strongly disagree" and "5 = strongly agree".

4. **Items of *social network***

Network size: how many people are important sources of knowledge on your new management practices? (0–10).

Structural holes: after listing the people you considered as important sources of information, you are asked to indicate whether or not these individuals know each other by circling "yes" in the matrix.

	1	2	3	4	5	6	7	8	9	10
1		Yes	Yes	Yes	Yes	Yes	Yes	Yes	Yes	Yes
2			Yes	Yes	Yes	Yes	Yes	Yes	Yes	Yes
3				Yes	Yes	Yes	Yes	Yes	Yes	Yes
4					Yes	Yes	Yes	Yes	Yes	Yes
5						Yes	Yes	Yes	Yes	Yes
6							Yes	Yes	Yes	Yes
7								Yes	Yes	Yes
8									Yes	Yes
9										Yes
10										

Link strength: Average link strength was operationalized based on interaction, affection and history of interaction. For each link you listed, you should answer the following three questions respectively to measure link strength.

(1) How frequently did you interact with each other during the past year?

1 = 2 times or less/year; 2 = 1 time/quarter; 3 = 1 time/month; 4 = 1 time/weak; 5 = 2 times or more/weak.

(2) You see him/her as a close friend;

1 = Strongly disagree; 2 = Disagree; 3 = Neutral; 4 = Agree; 5 = Strongly agree.

(3) How long have you been knowing him/her?

1 = within one year; 2 = one to three years; 3 = three to five years; 4 = five to ten years; 5=over ten years.

Items of network heterogeneity

(1) You two have different backgrounds;
(2) You two have different occupations;
(3) You two have different professions;
(4) You two are in different ages.

All items were measured on a 5-item scale, on which "1 = strongly disagree" and "5 = strongly agree".

5. **Items of *knowledge acquirement***

(1) New methods for internal operation could be obtained as soon as possible;
(2) You have various channels for new management practices;
(3) Knowledge from different channels is heterogeneous;
(4) The available innovation knowledge could satisfy the need for handling management problems;
(5) The obtained innovation knowledge is always reliable;
(6) You are able to acquire the latest knowledge on new practices;
(7) You can always get knowledge you need.

All items were measured on a 5-item scale, on which "1 = strongly disagree" and "5 = strongly agree".

6. **Items of *cognitive biases***

Overconfidence:

Please answer the following 10 questions with A and B selections, and then select confidence level for each, with "50% = complete uncertainty" to "100% = complete certainty"

Questions	Selections		Confidence level					
	A	B	50%	60%	70%	80%	90%	100%
Which year was Lean Production born?	1950	1970	1	2	3	4	5	6

(continued)

(continued)

Please answer the following 10 questions with A and B selections, and then select confidence level for each, with "50% = complete uncertainty" to "100% = complete certainty"								
Who was the first one proposing the concept of Learning Organization?	Peter Senge	Forrest	1	2	3	4	5	6
Where was Total Quality Management born?	America	Japan	1	2	3	4	5	6
How many cars were produced in China in 2009?	13 million	18 million	1	2	3	4	5	6
What percentage of enterprises is middle and small in China in 2009?	80%	99%	1	2	3	4	5	6
What is the position of Gross National Product of China in 2009 around the world?	3	5	1	2	3	4	5	6
What is the growth rate of Chinese economy in 2009?	9.1%	8.7%	1	2	3	4	5	6
What is the average distance between the Moon and the Earth?	384,400	538,400	1	2	3	4	5	6
How long is the Yellow River?	6422	5464	1	2	3	4	5	6
What is the total population of Shanghai by the end of 2009?	19 million	14 million	1	2	3	4	5	6

Items of illusion of control:

(1) Suppose the average success rate of adopting a new management practice is 50%, you believe that you are able to make it over 50%;
(2) Suppose the average cost of adopting a new management practice is 1 million, you believe that you are able to make it less than 1 million;
(3) Suppose that average 50% of employees would accept a new management practice, you believe that you are able to make it more than 50%;
(4) Suppose that average time for implementing a new management practice is 3 years, you believe that you are able to make it less than 3 years;
(5) Suppose that average success rate of a new management practice is 50%, you believe that you are able to make it more than 50%.

All items were measured on a 5-item scale, on which "1 = strongly disagree" and "5 = strongly agree".

Items of representativeness
Statistics indicated that CRM (Customer Relationship Management) was one of adoptive management innovation with highest failure rates? Would you adopt this new practice into your company when you hear that two other companies having close relationships with you have successfully implemented it?

It was measured on a 5-item scale, on which "1 = strongly disagree" and "5 = strongly agree".

7. **Items of *risk perception***

Suppose you are in charge of a large telecom operator, facing the challenges of both fierce competition press from the external environment and low efficiency of existing processes, you believe that:

(1) The overall risk of introducing the concept of Business Process Reengineering(BPR) to radically change the existing processes is high;
(2) The probability of failure is very high;
(3) The amount your company could lose is substantial;
(4) Introducing BPR will have a negative ramification for your company;
(5) There is a high probability of your company losing a great deal by introducing BPR.

All items were measured on a 5-item scale, on which "1 = strongly disagree" and "5 = strongly agree".

Appendix B
Questionnaire for Effects of Dynamic Capabilities

1. **Items (or indicators) of Adoptive Management Innovation**
   (1) We are able to renew various management practices or methods (number of new practices initiated);
   (2) We pay much attention to renewing management practices in improving performance;
   (3) We are able to identify management problems to launch an innovation;
   (4) We are able to initiate a management innovation before competitors (timing of initiation);
   (5) We are able to obtain an existing solution for management problems in time (deleted);
   (6) We are acquainted with emerging practices in market sources (for example, customers, suppliers, competitors, and consultants);
   (7) We are acquainted with new management practices from internal sources (i.e. employees)
   (8) We are acquainted with emerging practices in professional sources (for example, industry bodies, professional associations, and trade fairs);
   (9) We prefer to create a new proposal instead of imitating activities of other organizations;
   (10) We are able to reestablish a new proposal soon after outside search (deleted);
   (11) Newly reestablished proposals in our firm can always present the core concept of original management practices or method;
   (12) Newly reestablished proposals in our firm can always meet the demand of solving certain problems;
   (13) Nearly every innovation proposal set up during the past three years has been taken into practice;
   (14) Innovations in management practices are always implemented with support of the whole organization;
   (15) Innovations in management practices are always progressed as predicted;
   (16) Outputs of innovations can be produced as predicted to improve operation performance;

H. Lin, *Adoptive Management Innovation*,
https://doi.org/10.1007/978-981-10-7712-8

All items were measured on a 5-item scale, on which "1 = strongly disagree" and "5 = strongly agree".

2. **Items (or indicators) of Sensing capability**

    (1) Our managers are experienced and knowledgeable;
    (2) We have diversified paths for transferring information on external changes into the firm;
    (3) We have diversified paths for transferring information on internal changes up forward;
    (4) We are able to quickly select most important information;
    (5) We are able to quickly detect new opportunities and make reactions;

3. **Items (or indicators) of Absorptive capacity**

    (1) We are able to identify and acquire new and important knowledge from both internal and external environment;
    (2) We are able to analyze, process, interpret, and understand the information obtained;
    (3) We are able to combine existing knowledge with the newly acquired and assimilated;
    (4) We are able to create new information by incorporating the acquired;

4. **Items (or indicators) of Integrative capability**

    (1) We can systematically identify what resources that our firm can benefit from;
    (2) In our firm there is a system for transferring resources to various activities;
    (3) We own an effective communication and cooperation mechanism for vertical integration;
    (4) We own an effective communication and cooperation mechanism for horizontal integration;

5. **Items (or indicators) of Relational capability**

    (1) We have set up an effective external network where our firm is located in a central position;
    (2) We have set up a close relationship with the government and institutions;
    (3) We have set up various effectively cooperating relationships with other firms;
    (4) We have set up an effective internal network with all employees involved and actively interacted;
    (5) We have set up a trust-based internal network.

Kolibri 360 GmbH
Ferdinand-Jühlke-Straße 7
99095 Erfurt, Deutschland
produktsicherheit@kolibri360.de